W9-ATS-381

The Triangle Shirtwaist Fire and Sweatshop Reform in American History

Other titles in the *In American History* Series

Abraham Lincoln & the Emancipation Proclamation
0-7660-1456-8

The African-American Struggle for Legal Equality
0-7660-1415-0

The Battle of Gettysburg
0-7660-1455-X

Blazing the Wilderness Road with Daniel Boone
0-7660-1346-4

The Bombing of Pearl Harbor
0-7660-1126-7

The California Trail to Gold
0-7660-1347-2

Charles Lindbergh and the *Spirit of St. Louis*
0-7660-1683-8

The Chisholm Trail
0-7660-1345-6

The Confederacy and the Civil War
0-7660-1417-7

The Cuban Missile Crisis
0-7660-1414-2

The Dust Bowl and the Depression
0-7660-1838-5

The Harlem Renaissance
0-7660-1458-4

The Home Front During World War II
0-7660-1984-5

Jamestown, John Smith, and Pocahontas
0-7660-1842-3

The Jim Crow Laws and Racism
0-7660-1297-2

The Lindbergh Baby Kidnapping
0-7660-1299-9

The Little Rock School Desegregation Crisis
0-7660-1298-0

The Louisiana Purchase
0-7660-1301-4

The Manhattan Project and the Atomic Bomb
0-89490-879-0

The Mission Trails
0-7660-1349-9

The Natchez Trace Historic Trail
0-7660-1344-8

Nat Turner's Slave Rebellion
0-7660-1302-2

The New Deal and the Great Depression
0-7660-1421-5

The Pony Express
0-7660-1296-4

The Prohibition Era
0-7660-1840-7

The Santa Fe Trail
0-7660-1348-0

Shays' Rebellion and the Constitution
0-7660-1418-5

The Space Shuttle *Challenger* Disaster
0-7660-1419-3

The Transcontinental Railroad
0-89490-882-0

The Union and the Civil War
0-7660-1416-9

The Vietnam Antiwar Movement
0-7660-1295-6

Westward Expansion and Manifest Destiny
0-7660-1457-6

The Triangle Shirtwaist Fire and Sweatshop Reform in American History

Suzanne Lieurance

Enslow Publishers, Inc.
40 Industrial Road
Box 398
Berkeley Heights, NJ 07922
USA

PO Box 38
Aldershot
Hants GU12 6BP
UK

http://www.enslow.com

For Adrian

Library of Congress Cataloging-in-Publication Data

Lieurance, Suzanne.
The Triangle Shirtwaist fire and sweatshop reform in American history / Suzanne Lieurance.
p. cm. — (In American history)
Summary: Explores the people and events connected with the 1911 fire in a New York City sewing factory that killed 146 people and led to reforms in legislation regarding workplace safety.
Includes bibliographical references and index.
ISBN 0-7660-1839-3
1. Triangle Shirtwaist Company—Fire, 1911—Juvenile literature. 2. New York (N.Y.)—History—1898–1951—Juvenile literature. 3. Clothing factories—New York (State)—New York—Safety measures—History—20th century—Juvenile literature. 4. Labor laws and legislation—New York (State)—New York—History—20th century—Juvenile literature. 5. Sweatshops—United States—History—Juvenile literature. [1. Triangle Shirtwaist Company—Fire, 1911. 2. Industrial safety—History.] I. Title. II. Series.
F128.5 .L73 2003
974.7'1041—dc21

2002008761

Printed in the United States of America

10 9 8 7 6 5 4 3 2 1

To Our Readers: We have done our best to make sure all Internet Addresses in this book were active and appropriate when we went to press. However, the author and the publisher have no control over and assume no liability for the material available on those Internet sites or on other Web sites they may link to. Any comments or suggestions can be sent by e-mail to comments@enslow.com or to the address on the back cover.

Illustration Credits: Cornell University School of Industrial and Labor Relations Archives, p. 87; Enslow Publishers, Inc., p. 74; Photo by Byron, UNITE Archives, Kheel Center, Cornell University, Ithaca, NY 14853-3901, p. 45; Photo by Lerski, Barbara Wertheimer Papers, p. 84; Photo by Lewis Hine, UNITE Archives, Kheel Center, Cornell University, Ithaca, NY 14853-3901, p. 27; Reproduced from the *Dictionary of American Portraits*, published by Dover Publications, Inc., in 1967, p. 104; UNITE Archives, Kheel Center, Cornell University, Ithaca, NY 14853-3901, pp. 8, 10, 13, 15, 18, 21, 23, 34, 40, 53, 62, 65, 67, 69, 71, 75, 80, 81, 90, 92, 94, 111, 113.

Cover Illustration: UNITE Archives, Kheel Center, Cornell University, Ithaca, NY 14853-3901 (middle photo on right has been altered to accommodate layout).

★ Contents ★

Fire in the Factory

Saturday, March 25, 1911, was a bright, beautiful day in New York City. Workers in the Triangle Shirtwaist Factory did not have a chance to notice the nice spring weather, however. They were working overtime, busy sewing women's dresses and blouses called "shirt-waists." Their workplace was in the top three floors of the ten-story Asch Building. The Asch was a loft building, on the corner of Greene Street and Washington Place in Lower Manhattan, New York. Loft buildings were structures from six to twenty stories high, with an iron or steel frame. Each floor in a loft building was one huge room with high ceilings, which allowed for more space to store or display goods. But, in the top three floors of the Asch Building, wooden partitions had been built to make washrooms and changing rooms for the Triangle workers.

Several hundred immigrant girls and women, most between the ages of thirteen and twenty-three, worked quickly in the Triangle factory that day. They were try-ing to complete their assigned number of garments by

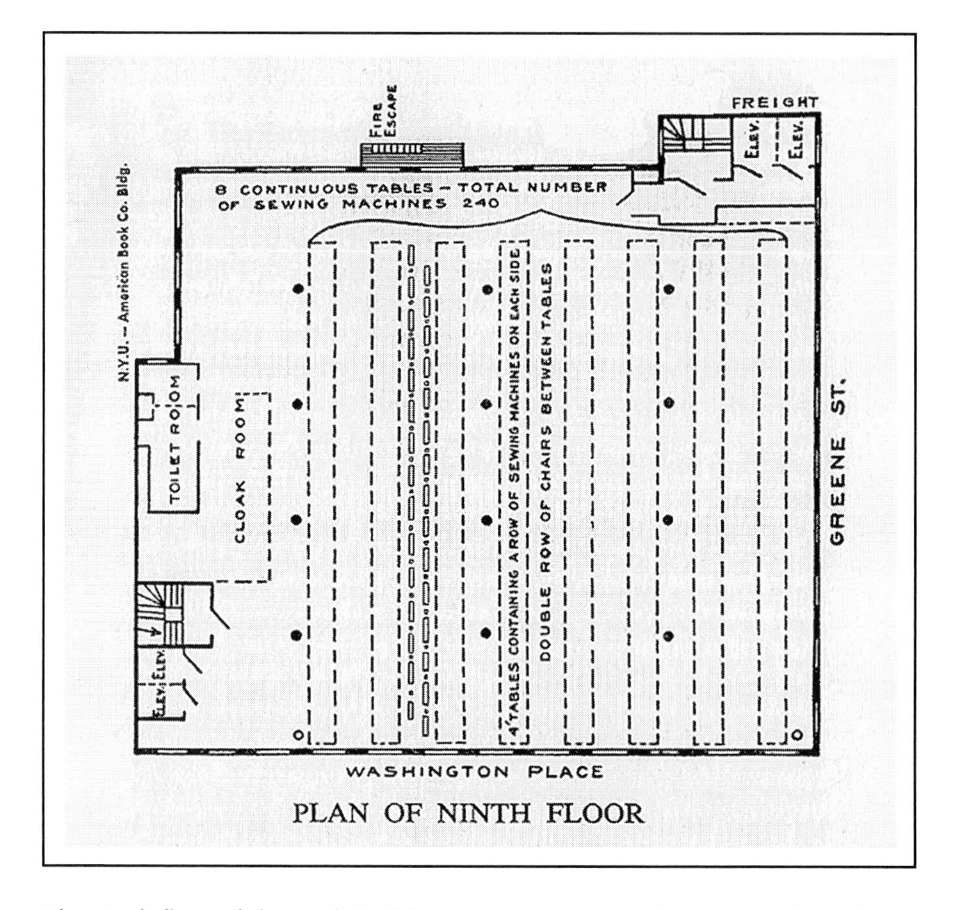

The ninth floor of the Asch Building was cramped with eight sewing tables (represented by long rectangles), holding 240 sewing machines, along with chairs for all the employees. This made it very hard for the workers to quickly make their way to the elevators, stairways, or fire escape.

quitting time. All the women could use the overtime money they would earn. Most of them made only a meager six dollars a week without it. Young women huddled over sewing machines at dark dusty workstations, amid stacks of highly flammable materials. Most of the women could barely speak English.

Power Off

Quitting time was 5:00 P.M. At 4:45 P.M. the workers were given the signal for "power off." It was time to shut down the sewing machines. Since it was Saturday, other companies in the Asch Building had let their employees go home early. The Triangle workers were the only ones in the building. The women went to collect their belongings to get ready to go home. They had no way of knowing that many of them would never make it out of the building alive.

A Fireproof Building?

The Asch Building was considered safer than most loft buildings. On the Washington Place side was a narrow stairway, about twenty-one inches wide. Also on this side of the building were two elevators, side by side. They were very small, though, only about five by five feet inside. Two other elevators were located on the Greene Street side of the building. These were used mainly for freight. A four-inch standpipe (a pipe in the walls, connected to a tank for holding water) with a hose was located on a rack on every stairway landing. There were also fire-pails on every floor (a total of

This narrow internal stairway was one of the few escape routes that the Asch Building had in case of fire.

twenty-seven pails). The building had no sprinkler system, but those were not required by law in New York factories at the time.

There had been many small fires in the loft before. Workers were always able to put them out. In fact, the Asch Building was considered "fireproof" by city fire inspectors because it was made mostly of stone. Only structures one hundred and fifty feet or taller were required to have metal trim, metal window frames, and stone or concrete floors. Since the Asch Building was only 135 feet high, its trim and window frames were made of wood. The floors, tables, chairs, and partitions in the Triangle factory were also made of wood.

Barrels of combustible sewing machine oil and bins of rags and scraps of flammable material were all around the workers.

Fire

As the workers scurried to put on their coats and leave for the day, Eva Harris, on the eighth floor, rushed over to the production manager, Samuel Bernstein. She said there was a fire between two tables on the Greene Street side. Some men were trying to put it out. Bernstein got two fire-pails full of water. He went over and tried to put the fire out himself.

The fire could not be put out, however. "It was in a rag bin and it jumped right up. Some of the women got pails and tried to help.

"But it was like there was kerosene in the water; it just seemed to spread it,"[1] said Bernstein. (Kerosene is a flammable substance.)

Water rarely puts out an oil fire and often spreads it. Many of the rags in the bin were probably soaked with oil that was used to lubricate the sewing machines. When workers poured water on these rags, this simply spread the fire.

Clothes hanging overhead on ropes caught on fire. Bernstein tried even more desperately to put out the fire, but the rope burned in half. The clothes fell onto people's heads.

One of the elevator men tried to help. Two other men tried to use the standpipe hose in the hall, but the valve would not turn. "It was rusted. And the hose, wherever it was folded, was rotten,"[2] one of them said later.

The fire was a "flash fire," meaning it spread quickly. Bernstein yelled to Louis Brown, a machinist, that they could not do anything. They needed to get the workers out—now!

By this time, the whole Greene Street side was on fire. The flames had begun to spread over the tables. The bookkeeper, Dinah Lifschitz, sent in an alarm, then telephoned upstairs to the office staff on the tenth floor. "I heard Mary Alter's [a cousin of owner Isaac Harris] voice on the other end. I told her there was a fire on the eighth floor, to tell Mr. Blanck. 'All right, all right,' she answered me."[3]

Max Blanck and Isaac Harris (right), owners of the Triangle Shirtwaist Company, escaped harm by climbing over to the New York University Building.

Isaac Harris and Max Blanck, the two men who owned the Triangle Shirtwaist Company, were working on the tenth floor. Blanck's two daughters, aged five and twelve, were with him that day. Their nanny had brought them into town for a shopping trip.

Harris pushed some women to the elevators on the Washington Place side. Then, Bernstein, who had run up the stairs from the eighth floor, led many of the workers as well as Blanck and his children to the roof. Some students in the adjoining New York University Building had just been dismissed from a tenth-floor classroom. They helped many Triangle workers climb over to the university building and escape safely.

A Human Chain

A group of three male workers made a human chain from the eighth-floor window of the Triangle building to the window in a building next door. A few women were able to cross over the men's backs to safety. Pauline Grossman was an eighteen year old who managed to escape from the building. She said, "As the people crossing upon the human bridge crowded more and more over the men's bodies, the weight upon the body of the center man became too great and his back was broken."[4] The man in the middle of the chain fell to the passageway below. The other two men lost their hold upon the windowsills. They fell, too, and the people who were crossing on this human bridge dropped eighty feet to their deaths with them.

Joseph Zito, an Asch Building elevator operator, heroically made as many trips as he could before the elevator finally broke upon reaching the ground floor for the last time.

As others screamed "Fire!" workers desperately tried to escape to the stairs or to the small elevators. There had never been a fire drill, so most of the women did not even know there was a fire escape. It went down the narrow vertical airshaft in the very center of the building and opened into a small courtyard.

Many workers rushed down the front stairs before the flames blocked them off. Others crammed into the tiny elevators, which were made to hold only ten people at a time. Women crowded in until there was no room or air left. Joseph Zito and Joe Gaspar, the elevator operators, made fifteen to twenty trips each bringing about twelve to fifteen women down to safety each time. Many of the women's clothes were still smoldering from the fire.

As Zito made his last trip down in the elevator, he heard bodies hitting the top of the car. People were jumping onto the top of the elevator since they could not get in. They were hoping to escape the fire and smoke as the elevator went down. Blood dripped on Zito and coins from the victims' pockets bounced through the shaft. By the time he got to the ground floor the elevator had stopped working, yet there were hundreds of workers still trapped upstairs.

No Way Out

Workers on the ninth floor had no warning before flames were everywhere. Fire leaped from underneath the worktables. It killed some women still at their workstations, bent over their sewing machines. Other workers jumped on the tables to escape, but their clothes caught fire and they died there, too.

For so many, there was simply no way out. Before the fire, foremen had locked at least two of the major exits to prevent workers from stealing scraps of cloth or taking rest breaks. Locking the exits like this was a common practice in many sweatshops. Other exit doors opened inward. As the panicked crowd pushed up against them, they became impossible to open. Young women huddled behind the exit doors. There was nowhere to go.

A few workers managed to find the rear fire escape. But, it was not very strong and it only reached to the second floor. As twenty workers climbed onto it, the fragile metal structure started to twist and come undone. The whole thing collapsed. All twenty people fell to their deaths.

To the Windows

Frantic workers pushed each other onto the factory floor trying to escape the flames. Dozens of women were shoved toward the windows. The flames from the floor beneath them were beating up in their faces. The women moved toward the ledges.

Down on the street, people began noticing smoke coming from the eighth floor. One man saw what appeared to be a bundle of cloth suddenly fly out the window. It hit the pavement. He thought Harris, one of the factory owners, was trying to save his best

The Asch Building's inadequate rear fire escape hangs in a twisted wreck.

SOURCE DOCUMENT

TRIES TO SAVE COMPANION.

THE CROWD YELLED TO THE TWO NOT TO JUMP. THE OLDER GIRL PLACED BOTH ARMS AROUND THE YOUNGER AND PULLED HER BACK ON THE LEDGE TOWARD THE BRICK WALL AND TRIED TO PRESS HER CLOSE TO THE WALL. BUT THE YOUNGER GIRL TWISTED HER HEAD AND SHOULDERS LOOSE FROM THE PROTECTING EMBRACE, TOOK A STEP OR TWO TO THE RIGHT AND JUMPED.

AFTER HER YOUNGER COMPANION HAD DIED THE GIRL WHO WAS LEFT STOOD BACK AGAINST THE WALL MOTIONLESS, AND FOR A MOMENT SHE HELD HER HANDS RIGID AGAINST HER THIGHS, HER HEAD TILTED UPWARD AND LOOKING TOWARD THE SKY. SMOKE BEGAN TO TRICKLE OUT OF THE BROKEN WINDOW A FEW INCHES TO HER LEFT. SHE BEGAN TO RAISE HER ARMS THEN AND MAKE SLOW GESTURES AS IF SHE WERE ADDRESSING A CROWD ABOVE HER. A TONGUE OF FLAME LICKED UP ALONG THE WINDOW SILL AND SINGED HER HAIR AND THEN OUT OF THE SMOKE WHICH WAS BEGINNING TO HIDE HER FROM VIEW SHE JUMPED, FEET FOREMOST, FALLING, WITHOUT TURNING, TO THE STREET. . . .[5]

The fire caused great panic among the workers trapped inside. They could not decide whether to stay inside a burning building, hoping to be rescued; or jump to a likely death below. Here, one girl tries to stop another from jumping before finally deciding to jump herself.

material. Out flew another apparent bundle of cloth. The man suddenly realized that these were not bundles of cloth. Frantic workers who could not escape from the smoke and fire were jumping to their deaths. Their bodies were crashing to the street below. It was later estimated that those who hit the ground struck it with a force almost one thousand times their weight.

According to a *New York World* reporter, onlookers cried out in horror, and "screaming men and women and boys and girls crowded out on the many window ledges and threw themselves into the streets far below. They jumped with their clothing ablaze. The hair of some of the girls streamed up flames as they leaped."[6]

Fire Engines Arrive

The fire raged for almost ten minutes before Engine Company 33 arrived. The fire engine was not able to do much good when it got to the scene. The stream of water from the firemen's hoses would only reach as far as the seventh floor. The flames poured from the windows. With nowhere to go, more women jumped from the building. Many tried to grab the top of the ladder on the fire engine as the firemen raised it. But, the ladder would only reach between the sixth and seventh floors. Many of the women hit the glass sidewalk cover, then crashed through to the basement. Their broken bodies were soon covered by water from the fire hoses.

Firemen try to spray water at the building with inadequate hoses. The water barely reached over the seventh floor while the fire raged on the three floors above.

William G. Shepherd, a United Press reporter who happened to be in Washington Square that day, witnessed the tragic event.

"I saw every feature of the tragedy visible from outside the building," he said. "I learned a new sound—a more horrible sound than description can picture. It was the thud of a speeding, living body on a stone sidewalk.

"Thud—dead, thud—dead, thud—dead, thud—dead. Sixty-two thud—deads."[7]

Women continued to jump. Some were holding hands, jumping in small groups. Firemen raised a life net and held it tight, but so many women jumped at once, they broke right through it.

According to Fire Captain Howard Ruch, "The force was so great it took the men off their feet. Trying to hold the nets, the men turned somersaults and some of them were catapulted right onto the net."[8]

Alfred K. Schwach, a student, saw four men try to catch some of the women in a horse blanket. "It gave way like paper as the girls struck it," he said.[9]

By this time thousands of people stood on the sidewalk, screaming in horror. There was nothing they could do.

Burned Alive

Up on the ninth floor the women could not jump. They were jammed into the windows and were being burned alive. A great mass of burning, smoking, flaming bodies came crashing to the sidewalk.

By the time other fire engines arrived, there were so many bodies on the street, the trucks had trouble getting through to the burning building. The smell of blood was also so strong that horses pulling the fire engines were frightened. They reared in panic. Several horses pulled free of their harnesses and bolted down the streets.

Over in Half an Hour

According to an estimate later made by Fire Chief Edward R. Croker, when the fire started there were approximately six hundred women and one hundred men at work in the Triangle factory. The fire did not

Crowds gather near the Asch Building, on the corner of Greene Street and Washington Place.

go below the eighth floor, but it was one of the most deadly fires New York had seen in many years, killing 146 workers. All but twenty-one of these were females, mostly around fourteen years old. The disaster lasted less than thirty minutes. No one ever found out how the fire started, but there were several theories. Supposedly, just as the first women started jumping from the building, many people who were on Greene Street that afternoon heard a small explosion and the sound of breaking glass coming from the Asch Building.

A man named Samuel Tauber, who had been employed as a foreman in the Triangle Company, told about a fire that had started on the eighth floor in 1909. According to *The New York Times*, Tauber explained that the motor that supplied power for the two hundred sewing and cutting machines on the eighth floor had emitted a flame. This flame set fire to some cuttings nearby. Tauber said that the fire had not been serious, but it had thrown the girls working there into a panic. He said he believed that this recent fire might have been caused in the same way.[10]

No matter what started the Triangle Shirtwaist fire in 1911, the conditions that led to it had been developing for years, and the effects of this horrible event would last forever.

2

AMERICA'S FACTORIES IN THE EARLY 1900S

During the nineteenth and early twentieth centuries, almost 50 million people in Europe left their homelands for better living and working conditions. The majority of these people headed for the United States. They heard America promised jobs, freedom, and the opportunity to make a fortune.

From 1824 to 1924, 34 million immigrants came to America.[1] A great number of these arrivals started in the mid-1840s, when Europe experienced a great famine. Most of these immigrants were Northern Europeans from Ireland, England, Germany, and Scandinavia. They were trying to escape starvation, unfair or changing governments, revolutions, and the chaotic social change caused by the Industrial Revolution. The next wave of immigrants occurred from 1890 to 1924. This time they came mostly from Southern and Eastern Europe. These people were trying to find relief from high taxes, poverty, and overpopulation. They were also victims of oppression and religious persecution.

Jews came from Romania, Poland, and Russia, where they had been driven from their homes by the government of the Czar (the emperor of Russia). Croats and Serbs, Poles, and Italians also left their homelands to escape one type of persecution or another.

New York City

Millions of these European immigrants settled in New York City, where most of them had entered the country. Most were poor and knew little about American life. They soon found out that the cost of living in this new country was far more expensive than in their old homelands.

In a 1904 book, Dr. Edward T. Devine, General Secretary of the Charity Organization Society of the City of New York, said that in New York City,

> where rentals and provisions are, perhaps, more expensive than in any other large city, for an average family of five persons the minimum income on which it is practicable to remain self-supporting, and to maintain any approach to a decent standard of living, is $600 per year.

He also said that considering an increase in the cost of living between 1900 and 1904, the estimate should probably be increased to seven hundred dollars if it were to apply to "the end of the period named" rather than to "average conditions of the past decade."[2]

By today's standards six hundred dollars a year does not seem like much. In the early 1900s, though, a family of new immigrants had a difficult time making only half that much in a year's time. Little children were important wage earners in families, too. A child of ten to twelve years old could find work finishing men's clothing or hemming dishtowels by machine. Young girls also helped women workers by sewing on buttons, pulling threads, or clipping threads between seams. Employers liked to hire children because they could generally pay them less money than they paid adults to do the same work.

In a speech to the National Consumers League in 1905, Annie S. Daniel reported on the average weekly

Workers in a small garment shop do sewing by hand.

income of men, women, and children employed in the garment business. She found that the average weekly income from the man's work was $3.81. Women earned, on average, $1.04 per week. The combined income of the men and women averaged $4.85 per week. Daniel said the additional sources of income came from the work of persons under eighteen years of age and from what could be received from boarders and lodgers. (Boarders and lodgers were people the families let live with them in exchange for money.) This made the average weekly income from all sources $5.69.[3] Even if a family could work all fifty-two weeks of the year, this still would only bring their yearly income to $295.88—way below the poverty level. More and more families crowded into a single apartment to save money on housing and other costs.

It is also important to note that whatever payment workers in the garment industry received for their work varied from shop to shop. Since there were no regulations regarding salary (or working conditions or safety requirements), employers tried to get by with paying workers the least amounts they could.

The Popularity of the Shirtwaist

One of the most popular garments for women at this time was the shirtwaist. The shirtwaist was a tailored, long-sleeved blouse with a high neck. Very soft fabrics were used for this garment.

The shirtwaist became popular because of Charles Dana Gibson. Gibson was an illustrator and his drawings appeared in a number of popular magazines such as *Scribner's, Harpers, Collier's,* and *The Century*. His illustrations set the fashion trend for American women at the turn of the century, creating "the Gibson Girl"—an attractive woman with her hair knotted on top of her head. She wore a long-sleeved shirtwaist blouse and a long skirt. Many young women wanted to be like the Gibson Girl, and the demand for shirtwaists grew. More and more shirtwaist factories opened to meet this demand.

Overcrowding

With rapid urban growth and few housing codes, severe overcrowding resulted in many American cities. New York City's Lower East Side became one of the most densely populated districts in the world.

Three-room apartments with just a living room, kitchen, and bedroom were often used as shops as well. In the early 1900s, six people might live in a shop, while as many as thirty people would also work there. If that were the case, no one had any privacy since every room was used for living, working, and sleeping space. The kitchen table was used as a worktable, and people often slept in shifts. Several families and workers shared the few outdoor toilets, and, later, indoor toilets that were located in hallways.

Finding Work

Most immigrants knew little or no English. They often had few skills that were desired by employers.

However, most of them had learned to sew at home, and thousands found employment in the sweatshops of the garment business. But, life in the sweatshops would be much harder than most of these people could ever imagine.

The Sweating System

Today, the word "sweatshop" usually means a place so hot and stuffy that workers "sweat" as they desperately try to make a living there. But, the term "sweating" actually means something quite different. According to an 1892 report of the Illinois Bureau of Labor Statistics, "sweating consists of the farming out by competing manufacturers as to competing contractors of the material for garments, which in turn is distributed among competing men and women to be made up."[4] In other words, a manufacturer would give work to a factory owner, and pay him a certain amount to get the work done. The factory owner would sell the work to other people. He would pay the workers as little as he could so he could keep most of the money he would receive from the manufacturer.

Under the sweating system the foreman acted as a subcontractor, with his own small shop. The subcontractor's main job was to organize and employ immigrants. Subcontractors were usually recent immigrants themselves. As such, they became organizers and employers of their fellow immigrants since they were often connected by the same language and religion. Some subcontractors were cruel bosses. They squeezed as much profit from their workers as they could, working them until they were worn out and almost ill. Other subcontractors were not much different from their employees. They were also victims of a very competitive market, yet they tried to be fair. The best subcontractors could easily persuade their neighbors or their wives and children to work for them in order to obtain the cheapest help.

The subcontractor could increase the number of people he employed at very short notice. Since the work doubled during the busy season, the number of people employed doubled, too. This constant change in the work force put additional strain on the workers. They never knew if they would be needed or let go.

Setting Up a Sweatshop

It was relatively easy and inexpensive to set up a sweatshop in 1900. All it took was a small amount of money, a couple of sewing machines, some tables and chairs, and a place to work. To compete against factories with modern equipment, subcontractors paid very low wages and located shops where rents were cheap. Even so, with extremely low profits most shops went out of business within a few years.

Typical Sweatshops

Sweatshops took many forms, but they were all places where the working conditions were awful, the hours were long, and the pay was low. Some sweatshops were housed in the basements of old buildings. Others occupied several floors of large lofts, much like the Triangle Shirtwaist Factory.

Regardless of where a sweatshop was located, it was a place where men, women, and children worked together, under the worst of conditions. There was little light and poor ventilation. Sweatshops were cold and damp in winter, hot and stuffy in summer. All year round the air was foul, as odors from nearby sewers filled the shops, along with the fumes of gasoline stoves and charcoal heaters. Cotton fibers from fabrics and threads drifted through the air, too. Often there were few restrooms, or no water supply at all. Sometimes workers had a dressing room where they could leave their coats and hats and purses. Most often they just hung these items on a hook beside their workstations.

Many workers were even required to furnish their own sewing machines. They carried them on their backs, to and from work each day. Inside the shop, machines and workstations were lined up, side by side, with hardly any room for movement. Although New York laws stated that every factory worker should have 250 cubic feet of air, most of this space was above the worker, since the loft buildings had high ceilings (ten or eleven feet). Most of the floor space was taken up by tables, workers, and machines.[5]

Gas lights provided dim light for the workers. As Clara Lemlich, a young worker at the Triangle Factory and a union activist, described,

> There is just one row of machines that the daylight ever gets to—that is the front row, nearest the window. The girls at all the other rows of machines back in the shops have to work by gaslight, by day as well as by night. Oh, yes, the shops keep the work going at night, too.[6]

Piles of scraps and fabric littered the floors. For many, the worst part of the sweatshops was the noise—the constant hum of the machines and bobbing needles as workers drove the machines as fast as they could to keep up with their bosses' demands.

The work was miserable. Still, it provided many new citizens with a way to make a living and save for a better future for themselves and their families. Some immigrants started working in small shops. Eventually, a few of these people owned large clothing firms themselves, though this was very rare. But others died from disease, malnutrition, accidents, and exhaustion before they could ever manage to get out of the sweatshops.

Pauline Newman was only twelve years old when she first came to the Triangle Shirtwaist Factory. She was assigned to a corner where workers as young as eight or nine years old trimmed threads from finished clothing. She later recalled, "We worked from 7:30 A.M. to 6:30 at night when it wasn't busy. When the season was on we worked till 9 o'clock. No overtime pay." She also remembered what it was like to work under such harsh conditions:

> They were the kind of employers who didn't recognize anyone working for them as a human being. You were not allowed to sing. Operators would like to have

Clara Lemlich, a sweatshop worker, decided to do something about the horrible conditions she was forced to work in and became a leading union activist.

> sung . . . and weren't allowed to sing. You were not allowed to talk to each other. Oh, no! They would sneak up behind you, and if you were found talking to your next colleague you were admonished. If you'd keep on, you'd be fired. If you went to the toilet, and you were there more than the forelady or foreman thought you should be, you were threatened to be laid off for a half a day, and sent home, and that meant, of course, no pay.[7]

There was no formal chain of command or hierarchy in most sweatshops since these workplaces were not regulated. Most often, a subcontractor or foreman supervised the workers. In the shirtwaist factories workers usually did only one specific job. For example, several women could be assigned to work on shirt cuffs, while others worked only on collars. Cutters were most often men, because they had the strength needed to cut through several layers of fabric. They would cut the fabric into the shapes needed for the garments. Children did the easiest jobs, like cutting off loose threads.

Work and Payment Systems

In the early days of America's sweatshops, Eastern Europeans introduced several work systems that actually worked against them. These systems kept the pay very low. One of these systems was the task system. Men and women worked as teams of sewing-machine operators, basters (those who sew something in place temporarily), and finishers with pressers and other helpers. Payment was for completion of a certain number of garments per day, instead of for so many hours work per day. Price cutting often led to the number of garments increasing over time and work-days extending far into the night. Many times a team would work fifteen to eighteen hours a day for six days, yet be paid for only four days' work.

Rose Cohen was just a child when she started work in a sweatshop on New York's Pelem Street. She soon learned how her employers would pay her less than she was worth. Her father explained, "It pays him better to employ you by the week. Don't you see, if you did piece work he would have to pay you as much as he pays a woman piece worker? But this way he gets almost as much work out of you for half the amount a woman is paid."[8]

Rose was not the only one who received far less than she was worth. In an April 15, 1911, article from *The Outlook*, Miriam Finn Scott explains how such unfair payment was commonplace at the Triangle shop:

> The Triangle firm had two systems of payment, piece-work and a fixed weekly wage, and it imposed upon each employee whichever method of payment it preferred. Becky was a swift and clever worker; in the busy season, she could make from eighteen to twenty dollars a week doing piecework. The Triangle Company, seeing how quick she was, with sharp business sense, changed her from piece-work to a weekly wage, and managed to get the same amount of work out of her for half the money.[9]

The reverse was practiced for slow workers. They were put upon piece-work.

Low Wages and Long Hours

Since major changes in the retail industry around this time expanded the market for clothes, there was intense competition among subcontractors for work. The development of large department stores and mail-order companies created an even greater demand for ready-to-wear clothing. With such intense competition, retailers pressured clothing producers to lower their prices. Clothing producers could do this because they would simply pay their workers less since new immigrants desperately needed employment. This kept wages down, yet workers worked long grueling hours.

According to a New York State law passed in 1899, all women and anyone under the age of eighteen could not work more than ten hours a day, or sixty hours a week. This law was revised in 1912. Under this new law, all women and anyone under the age of eighteen were limited to working nine hours a day, or fifty-four hours a week. Yet, employers continually broke these laws. When inspectors visited their factories, employers would hide the children or other minors in bins, or herd them to an elevator and then put it between floors until the inspectors were gone.

Mistreatment

Mary Van Kleeck was a noted social researcher and reformer. She studied women's employment at the Russell Sage Foundation in New York. She pioneered industrial investigation in the early 1900s. She found that workers were mistreated in a variety of ways. According to Van Kleeck,

> The makers of misses' and children's fine dresses work during the rush season from 8 A.M. until 9 P.M. with three-quarters of an hour for dinner [lunch] and one-half an hour for supper, five days in the week. Saturday until five, sixty-seven hours a week. Then they take work home, and toil until 11 o'clock at night.[10]

Van Kleeck also learned that during working hours the girls were not allowed to speak. They also suffered many other humiliations. "They were mean," said one girl. "Once I wanted to go home because my mother was very sick. The boss said, 'No favors here.' I was afraid to lose my position so I stayed."[11]

Safety and Health Concerns

Many factory owners in the early 1900s had little concern for safety or health issues. They were more interested in the bottom line (making a profit). In her 1905 speech to the National Consumers League, Annie S. Daniel also explained some of the health issues of the garment workers. She said,

> [Garment workers must] work on in dirt, often in filth unspeakable, in the presence of all contagious and other diseases.
>
> The sick as long as they can hold their heads up, must work to pay for the cost of their living. As soon as they are convalescent [beginning to recover] they must begin again. The other day a girl of eight years was dismissed from the diphtheria hospital after a

> severe attack of the disease. Almost immediately she was working . . . [12]

In the early 1900s, the law relating to manufacturing in tenement houses allowed the work for thirty-three distinct industries to be done in the living rooms of the workers. All of these industries required hand work or simple machinery. Daniel reported that she had seen clothing being made for infants and young children "on the same bed with children sick of contagious diseases and into these little garments is sewed some of the contagion."[13]

Those who did not become ill from the unsanitary working conditions often lost their lives or limbs in machinery accidents. There were no laws requiring owners to safeguard machinery and workers had to work fast and produce large quantities of garments. They often lacked the skills needed to operate machines. Lost fingers, scalpings, and spine malformations (caused from sitting stooped over for hours on end) were not unusual.

Sadie Frowne worked in a shop on Allen Street in Manhattan in 1902. She said,

> The machines go like mad all day because the faster you work the more money you get. Sometimes in my haste I get my finger caught and the needle goes right through it. It goes so quick, though, that it does not hurt much. I bind the finger up with a piece of cotton and go on working. We all have accidents like that.[14]

There were other dangers, too. It was reported by Frowne that,

> The dye from cheap cloth goods is sometimes poisonous to the skin; and the fluff from such goods inhaled by the operators is excessively irritating to the membranes, and gives rise to inflammations of the eye and various forms of catarrh [inflammation of a mucous membrane, especially of the nose and throat].[15]

Dr. George Price was one of the first tenement inspectors. He said that he felt the real menace to

Health and safety conditions for workers were not a top priority for factory owners in the early 1900s. It would take the devastation of the Triangle Shirtwaist fire to get city, state, and federal officials to stop and take notice of the danger that workers faced everyday.

workers' health and the real dangers in their lives were not in the nature of the garment industry itself. Rather, the real dangers were in the unsanitary conditions of the shops and their inadequate provisions for fire protection.

It took the tragedy of the Triangle fire to make people realize just how true Price's words were.

3

A General Strike Is Declared

A labor union is an association of workers that tries to improve the wages and working conditions of its members through group action. Today, workers in various occupations have their rights and interests protected by different labor unions. This has not always been the case.

Unions

The earliest types of labor unions in the United States were craft unions. A craft union organizes workers employed in the same occupation or craft. The earliest craft unions in the United States were formed shortly before 1800 in New York City and Philadelphia. These organizations represented such craftsmen as printers and shoemakers, whose work is considered "skilled labor."

Another type of union is the industrial union. This type of union organizes all workers in a particular industry, regardless of the workers' crafts. The United Steelworkers of America and the United Mineworkers of America are examples of industrial unions. Industrial unions are more commonly known as trade unions and these organizations represent "unskilled labor." In the early 1900s, many trade unions were being created for the first time. Today there is less of a distinction between craft unions and industrial unions.

A union can often negotiate higher wages, shorter hours, and better benefits (including health insurance and retirement plans) than individual workers can negotiate on their own.

Early Oppositions to Unions

The United States as a whole was not favorable to unions in the late 1800s and early 1900s.[1] Most of the garment workers were recent immigrants who desperately needed their income in order to eat. They could not afford to be out of work for several days or weeks for a strike. Also, since many of these immigrants spoke little or no English, it was hard for some of them to even understand what the strikers were trying to do. And, strikes sometimes became violent, which probably discouraged many from joining the union.

Employers often tried to prevent workers from joining the union. They locked the factory doors so union representatives could not get inside and speak to the workers. Workers who complained about this, or even mentioned the union, were called troublemakers and sometimes fired. Still, many workers realized that the union was their only hope for higher pay and better working conditions.

The Ladies' Waist Makers' Union

In August 1900, the Ladies' Waist Makers' Union was organized. It had only a handful of members and the union had not been able to negotiate with a single shop for better working conditions, safety, or wages. At this point, all the union could do was agitate (stir up interest and support for the union) and try to gain new members.

The few women who did join the union had a hard time. If they talked about the union to their fellow workers, or let anyone know they were members, they lost their jobs. Little by little, though, a handful of women began organizing. They led a series of small strikes and trained fellow workers what to expect from a strike and how to respond to the often violent tactics used by bosses to break the strike.

In the spring of 1905, the Ladies' Waist Makers' Union disbanded and reorganized as Local 25 of the International Ladies' Garment Workers' Union (ILGWU), which had been founded in 1900. Clara Lemlich and six of her coworkers at a local shirtwaist shop were on the executive board. They helped organize several small strikes.

The WTUL

The Women's Trade Union League (WTUL) played a key role in reforming women's working conditions in the early twentieth century, too. It helped women start unions in many industries in many cities. This organization also provided relief, publicity, and general

Members of the Womens' Trade Union League worked hard in their New York office.

assistance for women's unions. Margaret Dreier Robins was president of the WTUL from 1906 to 1922. Her sister, Mary Dreier, was president of the New York WTUL.

The WTUL would have a major role in supporting the shirtwaist strike of 1909. It would raise money for relief funds and bail (money given to the court for the temporary release of a person arrested). It would also revive a local chapter of the ILGWU, organize mass meetings and marches, and provide pickets and publicity.

A Series of Small Strikes

One of the small strikes that occurred around 1909 happened at the Triangle Shirtwaist Factory. Oddly enough, it was a subcontractor who started the strike. He was tired of having to work his employees so hard. He protested to the manager, saying he wanted to leave and take the workers with him. He was told to report to the cashier for his pay and get out. Since he was afraid of being attacked as he left the building, he asked that someone go with him to the elevator. Management refused and the man was dragged out. As he was being taken away, he shouted, "Will you stay at your machines and see a fellow worker treated this way?"[2] Four hundred workers turned off their machines and walked out with him.

At the time of this strike there were only five hundred members in the union and the trade was not organized to any degree. Secretary Schindler suggested the shop and the workers try to work things out, but the Triangle Company "locked out" the workers. This meant that they closed the shop, then said they did not need more workers. They only hired back workers who did not belong to the union.

In September 1909, Clara Lemlich was working at the Leiserson Factory. She and six of her coworkers went out on strike. During this strike Lemlich was arrested seventeen times. She had six ribs broken by police and company guards who used their clubs on her. Although she was badly injured, she went back to her post on the picket line after several days and tried to gain more strikers.

During both the Triangle strike and the Leiserson strike, two of the demands workers made were for adequate fire escapes and unlocked doors from the factories to the streets. These demands were not met at either shop. Both of these strikes were still important, however, because they created some momentum so workers could push for a general strike (a strike among all shirtwaist workers).

General Discontent

By 1909, there were some thirty thousand to forty thousand shirtwaist workers in New York City. About one in every four was a man. Many of the subcontractors were men, too. Most of the cutters were men, and men and boys also operated the machines and did the pressing. Girls and women did many of the other tasks, such as cutting off threads and examining and folding garments.

Both men and women shirtwaist workers were becoming more and more dissatisfied with working conditions and wages. The average wage at that time (with allowance for lost time in the dull or "off" season) was five dollars a week for women. Samplemakers earned about twelve dollars. Learners (those just learning their job) averaged three dollars, and about 30 percent of workers were always in the learner class. The average pay for those who had been working for awhile, which included about 45 percent of the women, was six or seven dollars. That left 25 percent who averaged more than seven dollars. It was possible, but did not happen very often, for a woman to earn twenty dollars a week in the busy season.[3]

Men's wages were higher. They generally averaged between sixteen to eighteen dollars a week. Many times men earned higher wages because they were given better work. Employers also claimed that men were paid more because they worked faster.

Many shops would keep only the best workers during the summer months and lay off all other workers during this time. This made life especially difficult for the laid off workers because they had to find other jobs for those few months, which was almost impossible.

Work hours were different for each shop. Generally, though, everyone worked even more hours. In rush season, workers often worked until nine (or later), seven days a week. The period of full employment lasted five months, from January to May. In June, July, and August there was practically no work. From September to Christmas there was only part-time work.

Mary Van Kleeck talked to some of the shirtwaist workers. One woman explained, "I used to make out by twelve dollars in the busy season, but in the dull only four. That is I got twelve dollars for about three weeks."[4]

In addition to the long hours, the constant nagging and extreme pettiness made the workers feel on edge. "The forelady drives you. If you fix a pin in your hair or your collar, before you know it there is a forelady saying to you, 'It isn't six o'clock yet. You have no right to fix your collar,'" said one girl.[5]

Another girl reported,

> In the busy season we worked overtime until half past eight and nine o'clock every night. They gave us fifteen cents for supper, that is us week workers, but supper was sitting by the machine and eating and working at the same time. They were awful mean about fines. At first they used to make us lose a half day if we came late, even five minutes.[6]

All in all, conditions and wages were not good for the thousands of shirtwaist makers. Only about one thousand of these workers were union members. A general meeting was planned for November 22, 1909, at Cooper Union Hall to see what could be done about working conditions, wages, and hours. Most of those who would attend this meeting did not belong to the union.

The Cooper Union Meeting

Cooper Union's Grand Hall was full on the evening of November 22, 1909. *Forverts*, a popular daily Yiddish-language newspaper among immigrants at the time, reported,

> Never have our workers shown such enthusiasm as at the Cooper Union meeting yesterday evening. At least 2,500 men and women who work in the Ladies Waist industry [were there]. They filled up the big hall of Cooper Union. And many, many hundreds blocked up the street outside of the hall because there was not enough room in the hall.

> In the hall, the masses were electrified with the enthusiastic speaking from the chairman and the moving speakers.[7]

Since Cooper Hall was overflowing with people, several other places—Beethoven Hall, Manhattan Lyceum, and Astoria Hall—were set up as extra meeting sites.

Samuel Gompers was president of the American Federation of Labor (AFL), the largest labor union in the United States in the early 1900s. He addressed the crowd at Cooper Union that evening.

"I have never declared a strike in my life," said Gompers. "I have done my share to prevent strikes, but there comes a time when not to strike is but to rivet the chains of slavery upon our wrists." Gompers continued his speech, asking workers to "stand together . . . "[8]

The crowd loved Gompers' words. His speech was met with thunderous applause.

The next speaker on the agenda was Jacob Panken. As he began to speak, an impatient woman from the audience asked to say a few words. People shouted out to her, telling her to get up on the platform. The woman was Clara Lemlich, a twenty-three-year-old shirtwaist worker, who had been working and organizing workers for eight years. Lemlich had been so badly beaten up by thugs during the Leiserson shop strike, that people had to lift her up to the stage. She said, "I move that we go on a general strike!"[9]

The crowd rose to its feet and cheered Lemlich.

A committee was then appointed at the Cooper Union meeting. The sixteen members of this committee went from one to the other of the overflow meetings, where the same motion to strike was offered and unanimously accepted.

The following morning thousands of shirtwaist workers went on strike.

The General Strike of 1909

On the first morning of the strike, November 23, Clara Lemlich spoke at fifteen union halls. One reporter wrote about the crowds of strikers:

> If you go down to the East Side these cold November days, you may see excited groups of women and girls standing at the street corners, gathered in public squares and crowded in the doorways. Go to the halls up and down Clinton and Forsythe streets and you will find similar groups multiplied till the overflow blocks the traffic . . . [10]

Out of forty-three thousand shirtwaist workers in New York City in 1909, more than twenty thousand went out on strike. The strike became known as the "Uprising of the 20,000."

Some of these girls who went out on strike felt they were receiving adequate wages. They were not displeased with the hours or the sanitary conditions of their shops. They struck to recognize the union and to express sympathy for the great majority of those who were underpaid, overworked, and housed in dark, unsafe, dirty buildings. Even Lemlich told a group of well-to-do women, "I did not strike because I myself was not getting enough, I struck because all the others should get enough. It was not for me, it was for the others."[11]

"It's because there are so many girls who can't make decent living wages that we had to strike. I can work unusually fast, and I make $12 a week during the busy season: but there are many who make only $4," said another striker.[12]

Strike Headquarters

Clinton Hall was used as general headquarters for the strike. Each day of the strike, strikers met, grouping off so members from each shop could talk together. They planned who would picket during every hour of the day. More experienced picketers would be grouped with the inexperienced girls to show them what to do.

The union set up payment of benefit money for the strikers, since they were not working to earn a salary when they were on strike. But there was an almost total absence of requests for any benefit money for this strike. One girl explained that many had only joined the union since the strike. Since the union had done so much for them by taking care of the strike, they wanted to "get along as far as we can without any money from them."[13]

Since many of the employers refused to pay their employees wages that were due to them when the strike began, money got really tight for many of

Women volunteer to walk the shirtwaist strike picket lines.

the strikers. The police were not much help to the strikers in getting them the wages owed to them by employers. The police often beat and arrested the strikers and broke up the picket lines.

The union also grew during this time. Within five days of the start of the strike, there were nineteen thousand new members.

Allies and Socialists

Although the strike was started by young women who were shirtwaist workers, many upper-class women participated in strike activities as well. These upper-class supporters were known as "allies." They offered financial and political support to the striking women. Among the allies was Anne Morgan, daughter of financier J. P. Morgan. She joined the WTUL and offered her help. "I have only known something of this strike for a short time . . . If we come to fully recognize these conditions, we can't live our own lives without doing something to help them, bringing them at least the support of public opinion," she said.[14]

Another ally was Alva Belmont who belonged to one of the four hundred richest families in New York. Belmont was the president of the Political Equality League. She raised money from personal appeals for funds to help the shirtwaist strikers. She criticized the shirtwaist manufacturers for using inside subcontractors.

Another group that offered support to the strikers was the Socialists. Socialists were members of political groups or parties who advocated more radical changes to the government. They wanted money to be more evenly distributed among members of society so there would be no rich and no poor. Although Socialists supported the strikers, they criticized the allies, believing they were insincere. Socialists felt that people such as Alva Belmont were involved in the strike only to gain support for the women's suffrage movement (to help women win the right to vote) and not to help the strikers.

Treatment of Picketers

When workers went on strike, the common practice would be to picket outside the doors of a shop at opening and closing hours. They would tell the "scabs" that a strike was on and ask them to come to union headquarters to learn about the union and the strike. Scabs were workers who chose not to strike and kept on working, or they were workers hired to fill positions vacated by strikers to try and stop the strike. Picketing was perfectly legal if it was done peaceably, but many of the picketers were arrested and fined in court without a hearing.

Every day the New York newspapers told of women who were beaten up while doing picket duty. The courts ignored the bleeding and bruised condition of the picketers and refused to stop the police who were responsible. Public opinion finally created a change in this attitude and the police became less harsh. On Friday, December 3, at noon, a parade of ten thousand girls, led by a committee who had personally suffered such violence, marched through the crowded streets. They went straight to Mayor George McClellan's office in City Hall and appealed to the mayor for protection against the police department. This seemed to have some effect.

"Steadily, surely, the tide of public interest and sympathy has risen. The city press has seemed to awaken slowly to the significance of the movement," reported Mary Clark Barnes in an article for *World Today*.[15]

In addition to the police, some shops hired thugs to make trouble for the strikers. These thugs would try all sorts of things to irritate or harass the picketers. They would get in front of one of the picketers and turn around suddenly every now and then, to block her way, or even run into her. They would knock off a striker's glasses and break them. This could be a real hardship if the striker could not afford new ones.

The Strikers Demands

Though the strikers had a difficult time with the police and hired thugs, they were not really asking for much from their employers. The strikers had a simple list of demands. They wanted a fifty-two-hour workweek, and not more than two hours overtime on any one day. They also wanted a closed shop (which meant no nonunion labor employed) and notice of slack work (meaning part-time work) in advance, if possible, or at least promptly on arrival in the morning. They were also tired of having to find other employment in the summer months. So, in slack season, they wanted employers to keep all workers on part time rather than a few workers on full time (as long as was possible). They also asked that all wages be paid by the firm. In other words, they wanted to get rid of the subcontractor system. They also asked for a wage scale that would be adjusted individually for each shop but the terms to be determined definitely in advance for all forms of work.

Source Document

The Uprising of the Twenty Thousands (Dedicated to the Waistmakers of 1909)

In the black of the winter of nineteen nine,
When we froze and bled on the picket line,
We showed the world that women could fight
And we rose and won with women's might.

Chorus:
Hail the waistmakers of nineteen nine,
Making their stand on the picket line,
Breaking the power of those who reign,
Pointing the way, smashing the chain.

And we gave new courage to the men
Who carried on in nineteen ten
And shoulder to shoulder we'll win through,
Led by the I.L.G.W.U.[16]

This song celebrates the waistmaker strike of 1909.

In addition to these demands regarding wages and working hours, the strikers wanted safer working conditions, including adequate fire escapes and unlocked doors from the factories to the streets.

Agreeing to Demands

Many of the smaller shops gave in right away and signed agreements with the union, but the real struggle went on for weeks. The wealthy and more powerful firms organized and formed an Association of Waist and Dress Manufacturers. The way they fought the union was by refusing to even recognize its existence. In mid-December of 1909, the association presented a contract to the strikers. They agreed to a fifty-two-hour workweek, piecework rate increases, the abolition of fines, and the improvement of sanitary conditions. The association still refused to recognize the union, however. The strikers rejected this open shop plan. An open shop would not recognize the union.

Constance D. Leupp, writing in the magazine *Survey*, said,

> So long as there are manufacturers in the trade who employ sweated labor, they can always underbid union shops. On the other hand, employers with the best intentions, who use both scab [(nonunion)] and union labor, will in a rush season make demands to which union members cannot accede and thus by degrees they must be driven out of the mixed shop.[17]

Morgan and other allies felt rejecting this plan was a poor decision on the part of the strikers. Morgan blamed the Socialist influence for the strikers' rejection. The public also did not understand why the workers would not give in on this one issue. But workers knew everything else could be withdrawn at some later date without protection provided by the collective action of a union.

The Strike Ends

The strike ended inconclusively in February. The union called off the strike since the strikers' demands had been granted in the majority of shops in New York. Also, many workers needed to get back to earning a living. Thousands of women returned to the shops. The strike was not totally successful. More than one hundred fifty large firms did not settle, and many refused to grant the demands of Local 25 of the ILGWU. The Triangle Shirtwaist Company was one of the shops that did not agree to shorter weeks, higher pay, safer working conditions, or recognition of the union. In fact, during the strike, Blanck and Harris, owners of the Triangle Shirtwaist Company found plenty of nonunion workers who crossed the picket line and were willing to work, so they refused to bargain with workers at all.

After the strike was over, the Triangle workers were not any better off than they had been before the strike. They still worked a fifty-nine-hour workweek and the Asch Building, where they worked, was still a firetrap.

The Cloakmakers' Strike of 1910

A year later, the cloakmakers' strike of 1910 ended with a historic agreement that established a grievance system (an agreed upon way to handle work-related complaints) in the garment industry. Still, many shop owners ignored basic workers' rights. They continued to impose unsafe working conditions on their employees. This new grievance system did not prevent shop owners from imposing unsafe working conditions on their employees.

It would take 146 grisly deaths before conditions for garment workers would change very much.

A City Mourns

Almost two years after the shirtwaist workers' strike of 1909, the Triangle Factory fire made many people keenly aware of the tragic consequences of inadequate fire escapes and locked doors. Everybody who witnessed the scene at the Asch Building would never forget what they saw.

After the Fire

After the fire was put out, a heap of corpses lay on the sidewalk for more than an hour. The firemen had been too busy dealing with the fire to see if there were any living among the dead. Finally, when the fire was out and the excitement had died down a bit, some of the firemen and policemen noticed something among the heap of bodies. About halfway down in a pile of dead girls, they found one girl who was still breathing. It was too late, however. She died two minutes after she was found.

Inside the Building

It was more than an hour and a half before the firemen could enter the eighth floor, where the fire had

FIRE TRAP VICTIMS BURIED

DRAFT NEW LAW TO SAVE SHOP WORK

THE WEATHER.

NEW YORK EVENING JOURNAL

8TH EDITION EXTRA

NEW YORK, TUESDAY, MARCH 28, 1911.

ONE CENT

SHEEHAN LOSES IN CAUCUS

WOMAN TELLS OF FIGHT FOR LIFE AT BARRED DOORS!

SURVIVOR'S REMARKABLE STORY.

7 HURT AS AUTO HITS AMBULANCE

POLICEMAN HURT BY RUNAWAY HORSE

ENDS HIS ILLS BY BULLET IN HEART

American Readers Will Take Your Rooms

SUES EX-HUSBAND TO GET SON AND $50,000

Americans read their newspapers every day to keep up with the newest stories from survivors of the fire. This headline appeared three days after the tragedy in the March 28, 1911, issue of the New York Evening Journal.

Source Document

They jumped, [they] crashed through broken glass, they crushed themselves to death on the sidewalk. Of those who stayed behind it is better to say nothing except what a veteran policeman said as he gazed at a headless and charred trunk on the Greene Street sidewalk hours after the worst cases had been taken out:

"I saw the Slocum disaster, but it was nothing to this."

"Is it a man or a woman?" asked the reporter.

"It's human, that's all you can tell," answered the policeman.

It was just a mass of ashes, with blood congealed on what had probably been the neck.[1]

The victims of the Triangle Shirtwaist fire were in horrible condition by the time they were found by policemen and reporters. In the dialogue above, the policeman refers to the "Slocum disaster," a fire on the ship the General Slocum *that occurred off the coast of New York City and claimed over one thousand lives.*

started. When they got there they saw fifty dead bodies on the floor.

In the elevator shaft was a pile of bodies. Firemen estimated there were twenty-five bodies of girls who had jumped down the elevator shaft after the elevator had made its last trip.

On the top two floors of the building the sights were even more gruesome. The floors were black with smoke. As the smoke cleared, firemen saw bodies burned to bare bone. Entire skeletons were bent over sewing machines, nothing but charred, headless trunks. Other piles of skeletons lay before every door and twenty-five burned bodies were found in a cloakroom.

Numbering the Dead

The bodies had to be gathered and numbered, to help with identification later. A policeman went around to each of the broken bodies with tags. He fastened a tag to the wrist of each of the dead women, then numbered each tag with a lead pencil. One of the dead women wore an engagement ring. Later, a total of eleven engagement rings would be found among the dead young women.[2]

United Press reporter William Shepherd said that a fireman who had come downstairs from the building told him there were at least fifty bodies on the seventh floor. This fireman must have meant the eighth floor. Fifty bodies were found on the eighth floor. No bodies were found on the seventh floor because the fire did not reach down there. Shepherd also said, "Another fireman told me that more girls had jumped down an air shaft in the rear of the building. I went back there, into the narrow court, and saw a heap of dead girls. . . . "[3]

Men attach identification tags to victims.

The remains of the dead would be taken to the morgue for identification. Many of the bodies were headless and so charred they hardly resembled human beings.

Identifying the Dead

By dusk that evening more than ten thousand people stood outside of the Asch Building as news of the fire had spread throughout the city. During the night, ambulances and police patrol wagons carried the dead to the Bellevue Hospital morgue on Twenty-seventh Street and to a temporary morgue at a tin-roofed pier on the East River. It took so long to remove all the bodies from the fire that patrol wagons used to transport the bodies to the morgues lined up on a side street. They looked like taxicabs waiting for riders. Then, they moved like a funeral procession. They traveled from Washington Square up Broadway to Fourteenth Street. Then, from Fourth Avenue on to Twenty-third, then east to Fifth Avenue and down to the foot of Twenty-sixth Street.

Police had sent for seventy-five to one hundred coffins from the morgue. There were only sixty-five available, so an order for more was sent to the carpenter shop of Metropolitan Hospital.

For several days, families of Triangle workers came looking for their loved ones. The lucky ones were reunited. Others had to identify the remains of family members. This was not easy for some, since thirty-nine bodies were burned beyond recognition. Relatives searched these charred remains for a familiar shoe, a bit of clothing, a ring, or anything else that might belong to their missing loved one.

Two thousand people waited at the morgue for the bodies to arrive. Coffins, each containing a body, were lined up. A temporary police station was set up at the pier. Forty policemen were assigned to help the grief-stricken relatives identify the dead. In the end, seven bodies could not be identified.

Officers carry the personal items of victims in baskets.

Survivors

More than an hour after the last of the women had jumped, policemen who had approached the building to gather up the bodies and stretch them out on the opposite side of Greene Street found signs of life. One girl, Bertha Weintrout, the last girl to leap from the ninth floor, was still breathing. Two or three dead bodies were piled beside her. The policemen heard the girl sigh as they moved those bodies away. A policeman yelled for a doctor. The girl was still bleeding and dripping wet, but she was rushed to St. Vincent's Hospital.

Hyman Meshell was taken from the building four hours after the fire started. He was found in the basement, frozen with fear and whimpering like an injured animal. He was crouched on top of a cable drum, in water up to his neck. His head was just below the floor of the elevator. The flesh of the palms of his hands had been torn from the bones when he slid down the steel cable in the elevator. His knuckles and forearms were full of glass splinters he got when he beat his way through the glass door of the elevator shaft. Meshell was also rushed to St. Vincent's Hospital.

Another worker who managed to escape was Rose Rosenfeld. On the day of the fire, Rosenfeld was working on the ninth floor. As the fire started, she stopped to consider what the executives were doing. She thought they would be safer so she went up to the tenth floor, where their offices were. She found Harris, Blanck, and other workers taking the freight elevator to the roof, so she followed them. Firefighters pulled her and the others to the roof of an adjacent building. "Girls in shirtwaists, which were aflame, went flying out of the building so that you saw these young women literally ablaze flying out of the windows," Rosenfeld said later.[4]

Another survivor was Bessie Gabrilowich. Gabrilowich was on the ninth floor. Since one of the women there had just become engaged, someone had brought a cake. Slices were being passed around.

Family and friends file past bodies at a temporary morgue, trying to find the remains of loved ones.

Gabrilowich was teaching a dance step to a fellow worker just as someone screamed. The flames were coming up from the cutting room on the floor below. Gabrilowich went to look for a cheap straw hat she had bought the day before, when she heard a foreman shout to her in Yiddish, "Bessie, save yourself." She looked across the room and saw her friend named Dora, looking frightened. But when Gabrilowich looked again, Dora was gone. She was one of those who jumped from the windows and died.

Gabrilowich somehow found her way to a staircase, covered her face with her purse, and ran to the street. The next day, she went back to the street outside the building. The bodies were lined up so friends and relatives might identify them. A newspaper photographer took a picture of Gabrilowich as she collapsed at the sight.[5]

Relief Work

Sunday, March 26, 1911, the day after the fire, the Executive Board of the Ladies' Waist and Dress Makers' Union, Local 25 of the ILGWU, met to plan relief work for the survivors and the families of the victims. Meeting with them were the executive board members of the WTUL. At this meeting, a relief committee was appointed. This committee would appeal for funds to use for immediate relief for victims and their families. Another committee was appointed to plan a funeral protest demonstration. These two boards also instructed the union's attorney to take steps necessary to prosecute Harris and Blanck, the owners of the Triangle Shirtwaist Factory.

That same day the committee drew up an appeal for funds. Women from the WTUL began visiting the homes of victims to see what temporary relief was needed. They would report back to the relief committee with this information. Money for the union's fund began to arrive by Monday morning. That same day,

The ninth floor of the Asch Building was completely gutted by the fire. Skeletal remains of sewing tables are strewn across this workroom.

temporary relief was given to many of the victims. The *Forverts* newspaper also started a relief fund on Monday. It was then decided that in order to avoid conflicts or duplications of relief efforts, a Joint Relief Committee should be formed. This happened on March 29.

The Joint Relief Committee was made up of representatives from the Women's Trade Union League, the Workmen's Circle, the Jewish Daily Forward, and the United Hebrew Trades. Over the next several months this committee allotted lump sums of money to the families of the victims. Its executive committee also distributed pensions every week, supervised and cared for the young workers and children placed in various institutions, and found work and suitable living arrangements for injured workers after they recuperated.

The Joint Relief Committee worked together with the American Red Cross. The Red Cross collected $100,000 for the benefit of the families of those who had died in the fire.[6] Many of the dead girls did not leave a single relative in this country. Their families were either in Russia or Italy, so money was sent directly to them.

The union committee and the Red Cross agreed on dividing relief efforts into two groups. The union would handle those cases in which the victim was a union member or surviving families had union members. The Red Cross would handle all nonunion cases. Both the union and the Red Cross would meet to decide who would be best to handle a particular case since there were often questions about whether or not a victim was a union member.

D. A. (victim's initials) was a surviving shirtwaist maker who was nineteen years old and a union member. This victim lost clothing in the fire and suffered from nervous shock. She was given twenty-five dollars for clothes, and five dollars for room rent, for a total thirty dollars. She was also sent to a German Home for two weeks' recreation, which cost thirty dollars.

J. C. (victim's initials) was thirty-five years old, and a union member who earned twelve dollars a week. She died in the fire. She left a husband, whose arm was slightly injured, and three children all under school age. Her family was paid $101 for funeral expenses, $34 for emergency relief, and $400 to help her husband start a small grocery business (at the advice of the wife's relatives). The total money her family received was $535.[7]

Triangle Company Back in Business

Harris and Blanck offered to pay one week's wages to the families of the dead workers. People were shocked and outraged. Was this all a person's life was worth?

Three days after the fire, Harris and Blanck put a notice in the trade papers. It let everyone know they were back in business: "NOTICE, THE TRIANGLE WAIST CO. beg to notify their customers that they are in good working order. HEADQUARTERS now at 9-11 University Place."[8]

Not much had changed, however. The day after Harris and Blanck opened at their new location, the Building Department of New York found that 9-11 University Place was not fireproof, and Harris and Blanck had already blocked the exit to the one fire escape with two rows of sewing machines.

Funeral Services and a Day of Mourning

For weeks following the fire, one funeral procession after another made its way through the city streets.

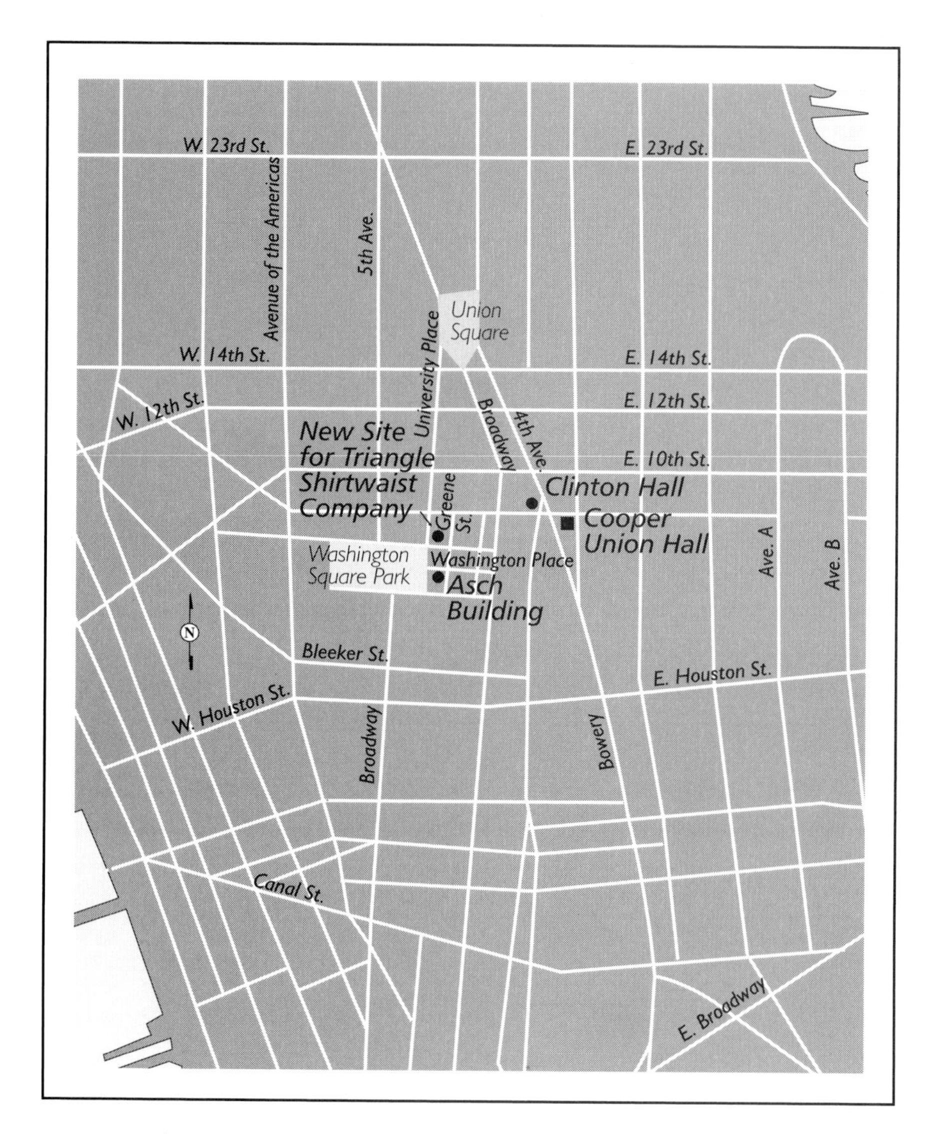

Workers from the Triangle Shirtwaist Company and other factories met at Cooper Union Hall, just a few blocks away from the Asch Building, and called for a strike in 1909. The strikers then used Clinton Hall as their home base for organizing picketers. After the deadly fire in 1911, Triangle Shirtwaist owners Max Blanck and Isaac Harris quickly reestablished their company in a new building at 9-11 University Place.

Through a special arrangement with the Workmen's Circle (the Jewish working class and death benefit society), the Jewish burials were made in the society's plot at Mount Zion Cemetery.

Several officials were against a single public funeral. They were afraid of how upset so many people could get over such a sad event. The International Ladies' Garment Workers' Union was not afraid. It proposed an official day of mourning with a mass funeral for the unknown victims and a march. On April 5, 1911, New Yorkers came out to pay their respects to the dead workers. All businesses and stores were closed for the day. Over one hundred twenty thousand people took part in the procession. An

Despite the rain, a large crowd gathers to honor victims of the fire.

estimated four hundred thousand watched from the sidewalks.

Martha Bensley Bruere wrote in *Life and Labor*,

> The fire is over, the girls are dead, and as I write, the procession in honor of the unidentified dead is moving by under my windows. . . . For two hours they have been going steadily by and the end is not yet in sight. There have been no carriages, no imposing marshals on horseback; just thousands and thousands of working men and women carrying the banners of their trades through the long three-mile tramp in the rain.[9]

As the crowd reached Washington Square and the Asch Building came into view, women began to express their deep sorrow.

Bruere writes, "It was one long-drawn-out, heart-piercing cry, the mingling of thousands of voices, a sort of human thunder in the elemental storm—a cry that was perhaps the most impressive expression of human grief ever heard in this city."[10]

Unidentified Are Buried

The coffins containing the remains of bodies no one had been able to identify were buried in Evergreen Cemetery. Hundreds of people waited in the heavy rain at the cemetery entrance. People were silent and respectful in honor of the dead, but after a while the crowd became curious and rowdy. The *Tribune* reported that as the hearses moved into the cemetery, boys ran through the grounds, jumping over and on the graves.

There were eight coffins. These were placed alongside a fifteen-foot-long pit that had been dug. Seven of the coffins had numbers, so they could be located later. The eighth coffin held "the dismembered fragments picked up at the fire by the police and unclaimed," reported the *Herald*.[11]

At one end of the pit was a small tent. As it rained, a small group of city officials huddled in front of the tent. First, Commissioner Drummond said a few words. Then, Monsignor William J. White read the Catholic service over one of the bodies in the coffins, as Father William B. Farrell made the responses.

The Reverend Dr. William B. Morrison read the Episcopal burial service over another body. Then, Rabbi Judah L. Magnus said some things in ancient Hebrew.

Nightmares of the Fire

Those who had witnessed the fire yet escaped physically unharmed were emotionally affected by what they had experienced. Rose Cohen was one young worker who had escaped the fire. She cried herself to sleep that night and for days afterwards.

The policemen and firefighters who responded to the Triangle fire were also deeply affected by the horrible sights and sounds they experienced. One firefighter searched the flooded basement, hoping to find survivors. Instead, he looked up to see two bodies, hanging from the overhead steam pipes. They had crashed through the Greene Street sidewalk and were mangled beyond recognition. It was a sight not easily forgotten.[12]

Placing Responsibility

Fire Commissioner Rhinelander Waldo said that the loss of lives in this tragedy had proved that while the Asch Building itself might have been considered fire-proof, it was evident that the contents of the building were not.

Ironically, just nine days before the fire, a local newspaper had run an article about the fire hazards that existed in most of the garment industry's sweatshops.

The public was outraged that something so horrible could be allowed to happen. They demanded that someone be held responsible. It would not be easy determining who was to blame for this tragedy, however.

Investigations and a Trial

Government agencies at the city, county, and state levels were feeling intense pressure to correct working conditions such as the ones that resulted in the loss of so many lives in the Triangle fire.

Special Meetings Held

Following the fire, the entire city of New York developed a new awareness of the conditions in its factories. Many rallies and other protests were held. Groups such as the Merchants' Association, the Public Safety Committee of the Federation of Women's Clubs, the Chamber of Commerce of New York, the Collegiate Equal Suffrage League, the Executive Committee of the Architectural League, and the Board of Directors of the United Cloak, Suit and Skirt Manufacturers of New York were among the groups that held special meetings during the week following the fire. They expressed a shared responsibility for the tragedy.

C. W. Phillips, assemblyman at the National Civic Federation's meeting, said that although New York was an industrial state with thousands of factories, it

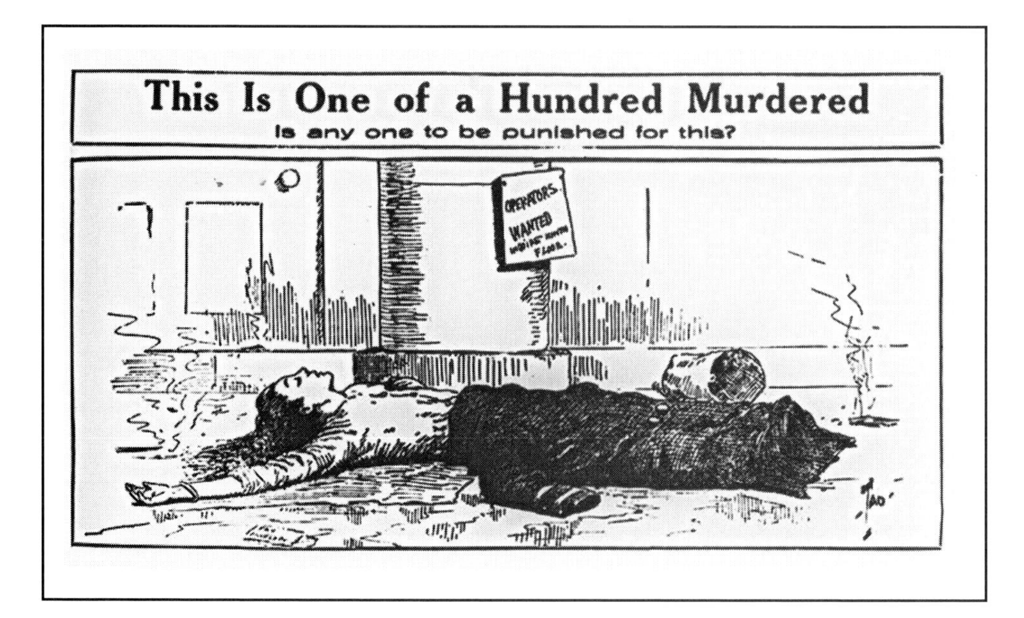

After the fire, many people felt that whoever was to blame for the fire should be punished.

"has 75 game protectors in its Department of Game, but only 50 human protectors in its Department of Labor."[1] ("Game" are animals that are hunted for food or sport.)

Dr. Anna Shaw spoke at a meeting of the Collegiate Equal Suffrage League. The League was formed to promote equality of the sexes under the law, and give women the right to vote. Dr. Shaw said, "There was a time when a woman worked in the home . . . all that has changed. Now she can no longer regulate her own conditions. She had been left . . . food for the flames."[2]

This political cartoon implies that city politicians were to blame for the fire, since fire codes and sweatshop laws were not strict enough.

Shifting the Blame

On March 27, city, county, and state officials met to discuss the issue of responsibility for the conditions existing in the Asch Building that led to the deaths of so many Triangle workers.

The condition of the fire escapes seemed to be one of the most important elements in determining responsibility. When the Building Department was charged with responsibility for the inadequate fire escapes at the Asch Building, Borough President George McAneny issued a statement saying that he felt the department was not responsible. He said there was not the slightest grounds for blaming Building Superintendent Rudolph P. Miller for this disaster. McAneny said the plans for the Asch Building were filed eleven years ago and were accepted as complying with the law. District Attorney Charles S. Whitman said the responsibility for fire protection in factories was with the State Labor Commission. State Labor Commissioner John Williams, however, stated that a recent court decision in 1903 had determined that the Building Department did, indeed, have control of the regulation of fire escapes in New York City.

It was clear that no one wanted to take responsibility for the inadequate fire escapes at the Asch Building. Legislation would later settle this issue. For the moment, it was unresolved.

Investigations continued as Fire Marshal William L. Beers had the Triangle owners and thirteen others give information he could use to determine the exact cause of the fire. According to the *New York Times*, Beers concluded that there was no explosion. A lighted match thrown into waste near oil cans or into clippings under cutting table number two, on the Greene Street side of the eighth floor, started the fire.[3]

A Meeting at the Metropolitan Opera House

Anne Morgan rented out the Metropolitan Opera House on behalf of the Women's Trade Union League for the evening of April 2, 1911. She wanted to have an open meeting that night, to bring people together toward a common goal of reform.

That evening poor East Side immigrants as well as wealthy members of society attended the meeting. Rose Schneiderman, a woman who had been a leader in the strike at the Triangle Company two years before, set the tone of the evening. She said this was not the first time girls had died in this city:

> Every week I must learn of the untimely death of one of my sister workers . . . the life of men and women is so cheap and property is so sacred! There are so many of us for one job, it matters little if 149-odd are burned to death . . . citizens, we are trying you now . . .[4]

That night, a resolution was made that called for the creation of a Bureau of Fire Prevention, and the addition of more fire and factory inspectors in the state. This would happen later, as the legislature passed the "Hoye bill," which was signed into law by the governor. As a first step toward this bureau, a twenty-five-member

Rose Schneiderman gave a rousing speech that inspired many of the attendees of the Women's Trade Union League meeting at the Metropolitan Opera House on April 2, 1911.

committee was selected that evening to improve safety in working places.

The Factory Investigating Commission

Public spirited citizens, representatives of the Fifth Avenue Association of the City of New York, the Committee on Safety of the City of New York, and other organizations presented the state legislature and the governor with information about the fire. This information revealed that many factories were unsafe, unhealthy, and lacked fire prevention measures or escapes. This, along with much public pressure, convinced the legislature that a full-scale investigation was needed so the legislature established the Factory Investigating Commission. The president of the Senate, Speaker of the Assembly, and governor appointed the nine-member commission. It would study issues related to the health and safety of workers, the condition of the buildings in which they worked, and existing and additional necessary laws and ordinances. Members of this commission included Senator Robert F. Wagner (chair), State Assemblyman Alfred (Al) Smith (vice-chair), and American Federation of Labor President Samuel Gompers.

Between 1911 and 1915, the legislature charged the commission with investigating sanitary and safety conditions, wages, living conditions of workers, and other related issues. The commission appointed directors for each investigation. Field agents were hired to carry out on-site inspections of factories and other work sites.

Investigation of factory conditions started in the fall of 1911. Dr. George M. Price was selected to direct the sanitation investigation. Price organized a group of field workers in September 1911, to carry out inspections. He said they would:

> compel the owners of the loft buildings to make radical improvements in their buildings, to spend huge sums for the protection of the lives and limbs of their tenants to make lessees of shops institute fire drills, buy fire extinguishing apparatus, and make other provisions for safety to rouse the workers themselves to the necessity of taking care of their own lives and health, something more than newspaper talk, than creation of safety committees or State commissions are necessary.[5]

The commission also chose H.F.J. Porter to direct field investigations related to fire hazards.

At the same time inspections were being carried out, members of the commission met in occasional executive sessions to make further plans, receive reports, and agree on recommendations. Questionnaires were given to businessmen, professionals, labor leaders, local government officials, engineers, fire department officers, and other persons. The questionnaires asked how the conditions of and the laws and ordinances regarding manufacturing could be improved. The commission received detailed comments from each person about these issues.

Francis Perkins was on the Factory Investigating Commission. She went on to be the first female cabinet member as President Franklin D. Roosevelt's secretary of labor from 1933 to 1945. In this photo, she discusses the Triangle Shirtwaist fire with a group of students at Cornell University over fifty years after her work on the commission.

In October 1911, the commission began conducting public hearings in New York City, Buffalo, Rochester, Syracuse, Utica, Schenectady, and Troy. Testimony was taken from hundreds of city and state officials, manufacturers, labor leaders, and working people. The commission used the information gathered from these hearings, along with what was collected through field investigations, questionnaires, and letters to draw conclusions about the working conditions in factories. They then developed recommendations as to how to improve poor conditions. The commission continued to hold hearings on issues under investigation until January 1915.

The commission and staff compiled all the information it acquired into several reports. The two main reports were "The Fire Hazard in Factory Buildings," and "Sanitation of Factories," published in the *Preliminary Report of the Factory Investigating Commission* in 1912.

The Commission's Preliminary Report

In its preliminary report to the state legislature, the commission outlined the range of its investigation. The report stated that the commission was

> charged with the duty of inquiring into the matters of hazard to life because of fire; danger to life and health because of unsanitary conditions; occupational diseases; proper and adequate inspection of factories and manufacturing establishments; manufacturing in tenement houses; the present statutes and ordinances that dealt with or related to the foregoing matters; and the extent to which the laws at that time were being enforced.[6]

The report went on to conclude that:

> A general awakening has taken place throughout the State. A far larger number of inspections by authorities have been made than ever before. No great reliance, however, can be placed upon such a momentary or spasmodic awakening. When its cause is no longer present, conditions relapse into their former state, and there is little real improvement.

> To improve the industrial situation permanently, clear, concise and comprehensive legislation is needed.[7]

Members of the commission would work hard during the next four years to see that such legislation was enacted.

The Trial

Meanwhile, soon after the fire, family members and friends of the Triangle fire victims said they would either start a petition or write personal letters to District Attorney Whitman, asking him to bring Harris and Blanck to trial. The two were indicted for manslaughter in April 1911.

Isaac Harris and Max Blanck went on trial for manslaughter for the death of one worker, Margaret Schwartz, on December 4, 1911. There seemed to be a strong case against them. Their attorney was Max D. Steuer. The case was prosecuted by assistant district attorneys Charles S. Bostwick and J. Robert Rubin. The trial lasted eighteen days.

More than one hundred fifty witnesses were called to testify about the fire, whether the doors were locked, if the owners knew they were locked, and whether the locked doors led to the death of Margaret Schwartz. Kate Alterman, another worker at the Triangle Company the day of the fire, told how she and Schwartz had tried to open the door to get out, but could not:

Assistant district attorneys J. Robert Rubin (middle) and Charles S. Bostwick (right) investigated the roof of the Asch Building with Coroner Holzhauser (left) in preparation for the trial of Isaac Harris and Max Blanck.

> I noticed someone, a whole crowd around the door and I saw Bernstein, the manager's brother trying to open the door, and there was Margaret near him. Bernstein tried the door, he couldn't open it and then Margaret began to open the door. I take [sic] her on one side—I pushed her on the side and I said, "wait, I'll open that door," I tried, pulled the handle in and out, all ways and I couldn't open it."

Alterman then described how she saw Schwartz bending down on her knees. Her hair was loose, and the trail of her dress was a little far from her, and then "a big smoke came and I couldn't see. I just know it was Margaret, and I said, 'Margaret' and she didn't reply."[8]

Alterman escaped by covering herself with dresses and a coat, then she leaped through the flames and firemen rescued her. Schwartz was not so lucky.

Workers testified that usually the only way out of the Asch Building at quitting time was through an opening on the Greene Street side, where the women's purses were inspected to prevent stealing. Worker after worker testified that he or she could not open the doors to the only escape route (the stairs to the Washington Place exit), because the stairs on the Greene Street side were completely covered by flames.

A parade of witnesses described the events of the fire in dramatic testimony. Before the jury went out to decide whether Harris and Blanck were guilty, Judge Thomas C. T. Crain instructed them that the key to the case was whether Harris and Blanck knew that the door was locked. Judge Crain told the jury:

During the trial of Harris and Blanck, the coroner and jury members questioned surviving employees to try to determine what exactly happened inside the building during the fire. Jury members needed as much information as possible before they could decide whether the owners of the Triangle Shirtwaist Company were at fault.

> Because they are charged with a felony, I charge you that before you find these defendants guilty of manslaughter in the first degree, you must find this door was locked. If it was locked, and locked with the knowledge of the defendants, you must also find beyond a reasonable doubt that such locking caused the death of Margaret Schwartz. If these men were charged with a misdemeanor I might charge you that they need have no knowledge that the door was locked, but I think that in this case it is proper for me to charge that they must have personal knowledge of the fact that it was locked.[9]

The all-male jury retired to deliberate at about 2:55 P.M. The jury returned to the courtroom at 4:45 P.M., less than two hours later. They rendered a verdict of not guilty.

Defense attorney Max Steuer was clever. He planted enough doubt in the jurors' minds to win a not guilty verdict. One of the jurors, Victor Steinman, was bothered by his verdict, however. He told a reporter from the *New York Evening Mail*,

> I believed that the Washington Place door, on which the district attorney said the whole case hinged, was locked at the time of the fire. But I could not make myself feel certain that Harris and Blanck knew that it was locked. And so, because the judge had charged us that we could not find them guilty unless we believed that they knew the door was locked then, I didn't know what to do.[10]

Steinman went on to explain that he felt the factory inspectors' duties were clearly outlined by the law. It was up to them, more than to Harris and

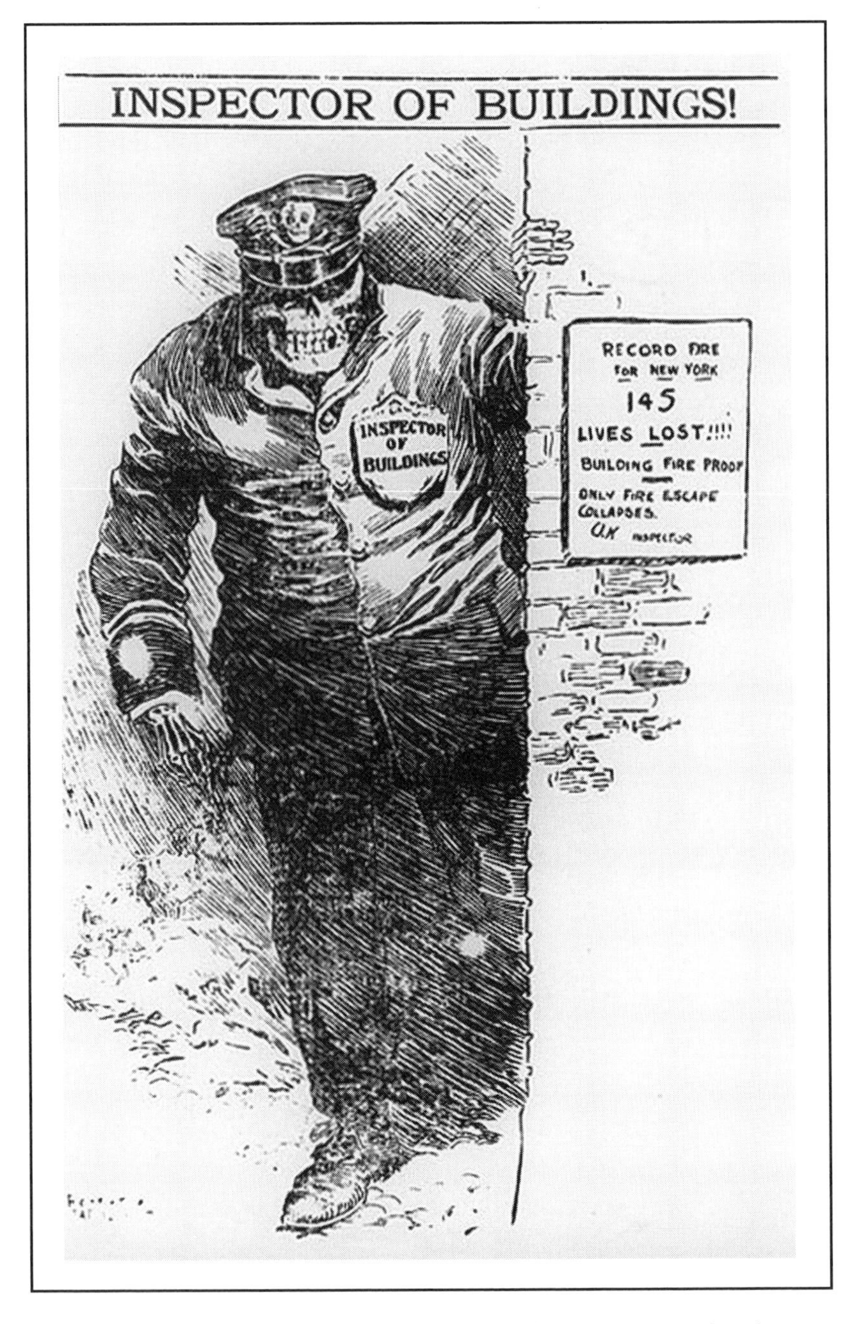

Some political cartoonists blamed building inspectors for the tragedy at the Asch Building. Here, a building inspector is depicted as a grim reaper.

Blanck, to see that the door was not locked. Steinman had seen the piece of charred wood and the lock with the shot bolt that the State put into evidence. But he also believed the testimony that the key was usually in the door and that it was tied to it with a piece of string. He said he felt that during the first rush for the door, some panic-stricken girl might have turned the key, trying to open the door. And, if that was the case, then Harris and Blanck could not have known about it, as the judge demanded they should, to be convicted.

Another juror said he felt the women probably panicked, causing their own deaths.

The prosecution demanded a retrial. The court denied it. Harris and Blanck went free.

Grieving families and much of the public were outraged. "Justice!" they cried. "Where is justice?"[11]

The *New York Sun* and the *New York Herald* reported that it was a fair trial and agreed that the verdict was not surprising, considering the contradictory evidence that was presented. The *New York Press* said, "The blood of those victims was on more than two heads, on more than twenty heads, perhaps on more than a million heads."[12]

Twenty-three individual civil suits were brought against Harris and Blanck. On March 11, 1913, three years after the fire, the two men settled. They were ordered to pay damages of just seventy-five dollars to each of the twenty-three families who had sued them.

6

Steps to Improve Working Conditions

The state's new Factory Investigating Commission took its job very seriously. Within its first year, the commission inspected 1,836 industrial establishments in New York and heard a total of 222 witnesses. Throughout this process, it also held hearings before the New York legislature and proposed new laws or amendments. The legislature in turn enacted legislation to correct past problems in the workplace.

Concerning the Asch Building at the time of the Triangle fire, the commission felt there was no question that the emergency exits from the building were inadequate. Yet, Fire Marshal Beers said, "I can show you 150 loft buildings far worse than this one."[1]

The commission found that at least fourteen industrial buildings in New York City had no fire escapes at all. It also found that the crowding on floors in the Triangle factory contributed to the number of lives lost in the fire there. Fire Chief Croker said, "The overcrowding of these loft buildings is a menace to life. . . ."[2]

Eventually, a series of corrective acts was passed. These laws were very specific. They stated that:

1. In factories there must be two exits per floor, one of these a staircase and another an interior or exterior enclosed fire escape.
2. If the area of the floor exceeds five thousand square feet, an extra exit is required (and for every additional five thousand square feet beyond this number, another exit is ordered).
3. If the building's height is over one hundred feet, there has to be at least one exterior enclosed fire escape accessible from every point in the building.
4. All stairways must be fireproof (concrete or brick).
5. All fire escapes must be iron or steel, and if enclosed, they must be enclosed by fireproof walls.
6. The number of workers allowed to work in factories is limited to the number able to safely escape from the building.

There were other new laws, too. During the Triangle fire, stacks of fabric and paper cuttings that covered the floor and tables fed the spread of the blaze. A new law required that all waste in factories, including cuttings of fabrics, be deposited into fireproof receptacles and that no waste be allowed to accumulate on the floor. The commission also recommended that smoking in all factories or manufacturing establishments be prohibited, and a notice to that effect (detailing the penalty for violations) be posted on every floor of such establishments in English and such other language or languages as the local fire commissioner or fire marshal directed.

Windows of wired glass were also recommended. The commission stated that all windows and doors leading to outside fire escapes shall be not less than two feet in width by five feet in height, and shall be constructed of wired glass. Wired glass helps keep objects from smashing through the glass and the wire holds pieces of broken glass together. By holding together, wired glass can also protect against break-in and the spreading of fire.

The commission also learned that many times exits to outside fire escapes and to interior stairways were unknown to many workers. Therefore, they felt it was necessary to indicate clearly the location of these exits. Also, clear and unobstructed passageways to exits should be insisted upon.

Thirty bodies were discovered in the shirtwaist company's open elevator shafts after the fire. In July 1911, the New York Legislature required that all elevator shafts in all city buildings be enclosed.

In 1912, legislation was enacted requiring the installation of an automatic sprinkler system in factory buildings over seven stories high with more than two hundred people employed above the seventh floor.

Fire Chief John Kenlon had previously reported to the commission that although an automatic sprinkler system would have cost the Asch Building five thousand dollars, he believed that no one would have died in the Triangle fire if an automatic sprinkler system had been installed.

It was also agreed that the lack of a fire drill at the Triangle factory caused panic when the fire broke out. Where fire swept through the building without warning, as it did in the Triangle factory, a fire alarm would have insured an earlier detection of fire and an earlier escape. Because of this, an addition to the labor law required a fire drill at least every three months, and the installation of a fire alarm signal system in any factory building over two stories high and employing twenty-five persons above the ground floor.

Hazards unrelated to fire safety were also discovered during the Factory Investigating Commission's inquiries. The commission found working children, lead poisoning, industrial accidents, insufficient ventilation, and not enough toilets in factories. As a result, child labor reforms were passed limiting the number of work hours for minors and prohibiting the operation of dangerous machinery by anyone under the age of sixteen. All industrial accidents and poisonings were required to be reported to the state. And, "suitable and proper" ventilation and washrooms were also required by law.

In order to enforce these new regulations, the legislature increased the number of inspectors from

47 to 125. The legislature also spelled out the labor department's duties and powers, and how inspections should be run. It also specified the duties and powers of the chief factory inspector and the first deputy inspector. Another part of the law outlined the fire marshal's duties in regard to supervision of adequate fire exits and fire drills.

In 1913, the fifty-four-hour workweek became law in New York State. By 1914, thirty-six new laws had been passed, including a women's minimum wage law.

Al Smith

Many of the new safety regulations presented in the years immediately after the Triangle fire probably would not have passed in the legislature without the support and hard work of State Assemblyman Al Smith. Many regard Smith as the hero of the reform movement. Smith was first elected to the New York Legislature in 1904. "He taught himself state government by reading every bill he could read, night after night," said writer Robert Caro.[3] Smith was very serious about his work with the Factory Investigating Commission. He never missed any of the factory inspection trips. When members of the commission were developing recommendations for what should be done to improve safety in factories, Smith told them he would fight to see that their recommendations became law.

Al Smith's campaign for the state legislature had been directed by the main Democratic political organization in New York City, Tammany Hall. Tammany Hall had been around since the late 1700s. It was infamous for its trading of jobs and political favors for money. One of its most famous leaders, "Boss" Tweed, had been jailed for cheating the city out of several million dollars. Most of the favors of Tammany Hall went to its Irish Catholic supporters. They had been in the first big wave of immigrants from Europe in the 1800s. But now in 1911, the second wave had come from Eastern Europe. Charles Murphy, Tammany Hall's leader, came to realize after the fire that his new constituents were the Eastern European immigrants, Jewish people from Russia and Germany, Italians, and others.

The reform, progressive, and socialist movements had come from these people. They had been trying for years to get legislation passed. Charles Murphy had recommended Al Smith to be put on the state Factory Investigating Commission. Now that Tammany Hall had accepted the new wave of immigrants, Al Smith worked for reform laws for all people.

The reforms that the New York Legislature passed as a result of the work of the Factory Investigating Commission were law in New York State only. Since then, the federal government has also dealt with workplace safety many times.

Roosevelt's New Deal

Many years after the fire, President Franklin D. Roosevelt created the New Deal. The country was suffering from the Great Depression, a time when many people were without jobs or any savings. Across the country, people took whatever jobs they could find. The New Deal was designed to help the American economy through a series of innovative regulations.

As part of the New Deal, Congress passed the Wagner Act of 1935. This act created the National Labor Relations Board and required private employers to deal with unions and not discriminate against union members. The act guaranteed workers the right to collective bargaining. Collective bargaining is negotiation between unionized workers and their employer. The aim of this bargaining is to reach an agreement on wages, fringe benefits, hours, and working conditions. The National Labor Relations Board also oversaw union elections and the settlement of labor disputes. The worst of sweatshop conditions began to disappear thanks to union gains backed up by government support.

Temporary Victory

The August 1, 1938, issue of *Life* magazine declared the war on sweatshops was won: "Thirty years ago the industry stank of the sweatshop and the cruelest kind of exploitation . . . Still numerous in 1933, the sweatshop is virtually gone today."[4] This was good news. But it did not last long.

Changes in the WTUL

After the Triangle fire, the Women's Trade Union League had evolved from an affluent women's group to a working class organization. It suffered financial setbacks throughout the depression and finally dissolved in 1950.

Things Start to Change

When Dwight Eisenhower was president in 1957, the Wages and Hours Division of the Department of Labor had one investigator for every 46,000 workers. Apparently, this was an adequate number because there was an overall decline in the number of abuses of the labor laws.[5] Sweatshops began to reemerge in the late 1960s. Changes in the retail industry, a growing global economy, increased reliance on contracting, and a large group of available immigrant workers in the United States all contributed to the sweatshop's reappearance.

Occupational Safety and Health Administration

The federal government has continued to make safety in the workplace an important matter, however. The Occupational Safety and Health Administration (OSHA) was created when Congress passed the Occupational Safety and Health Act. This act was signed into law on December 29, 1970, by President Richard M. Nixon. The law was designed to help prevent work-related illnesses, injuries, and deaths.

Under President Richard M. Nixon, the Occupational Safety and Health Administration was created in 1971. Though the agency has since reported success in preventing work-related injuries and deaths, it has also been criticized by both business and labor leaders.

According to the OSHA Web site, workplace deaths have been cut in half since the law was passed. The site says injuries have declined 40 percent since 1971. In 2001, this federal agency had a budget of $426 million, and 1,170 inspectors. Twenty-six of the states have similar programs and an additional 1,275 inspectors.

Sweatshops Reappear

After the mid-1970s, the number of investigators declined due to successive budget cuts during that time. When Ronald Reagan was president from 1981 to 1989, the investigators-to-employees ratio was only one per 110,000. After President George H. W. Bush left office in 1993, the ratio was one to 130,000. This trend did not change when Bill Clinton took office. By 1996, the ratio was then one per 150,000, for a total of only 781 investigators (unlike the 1,100 during Eisenhower's presidency).[6] These investigators could not possibly keep up with enforcing the laws. Sweatshops began to reappear.

UNITE

In 1995, the one-hundred-twenty-five-thousand-member ILGWU merged with the one-hundred seventy-five-thousand-member Amalgamated Clothing and Textile Workers' Union to form the Union of Needletrades, Industrial and Textile Employees (UNITE). This organization faces many of the same problems that earlier unions did.

Sweatshops Today

Sweatshops in the United States produce garments for the domestic market, mostly items that have short delivery times. These garments look the same as clothes produced in legal shops and they can be found in all kinds of stores, from discount shops to the finest boutiques.

According to Robert J. S. Ross, director of the International Studies Stream program at Clark University in Worcester, Massachusetts, six out of ten small contractor shops where workers sew clothing persistently break the labor laws. They fail to pay minimum wages or overtime. Employers are able to get away with breaking the laws because there are not enough investigators to see that the laws are enforced. Plus, sweatshops are often mobile, which makes them very difficult to regulate. "The equipment is really just a few sewing machines," said Ginny Coughlin of UNITE. "Just rent space, pay the electric bill, and you're in business."[7]

Undercover in a Sweatshop

In 1995, *Dateline* (a TV news show) producer Minnie Roh conducted an eight-month undercover investigation to explore the working conditions of garment factories. She wore a hidden camera and kept a journal of what she found. She found that situations in sweatshops are not that much different than they were in 1911.

Roh's first job was at a small shop on Long Island, New York. She worked all day making sweatshirts. As she explained,

> I have made 53 sweats this day. I have worked 8 hrs. . . . I made $3.71. And I am no means, by far, the slowest of the bunch. I timed myself. I can make a sweatshirt almost as fast as the other workers. Nancy had told me my work load was 150 sweatshirts. After working your hands to the bone, making 150 sweats, I would make $10.50. Minimum wage that would be a little over 2 hrs. work. But to make 150 sweats requires more than 2 hrs. of work.

At the end of the day, Roh said,

> So my first day in retrospect was an eye opener. The 1st words I uttered after entering the van was "Oh my God." I was just so happy. Happy to know that my day was over. But for the others? It was still not over. I will go back tomorrow. I must go back tomorrow.

> How many sweats I can turn around. Think about that I made $3.71 for 8 hrs. of work today.

A few days later, she moved on to another shop. While she was there she encountered dangers not unlike those in the Triangle factory so many years ago. In the diary Roh kept of her experiences she says,

> Boom! Spark! An electrical wire bursts right in front of me and all this smoke comes billowing out. At first, I thought, oh how amazing! Got to get a shot of that. Then, I thought, well that's pretty . . . dangerous! But everyone just waved the smoke away, just kept on working . . . Later on in the day, I canvassed the factory and the only other exit is a freight elevator that was padlocked. There were wires all over the place, dangling in front of my face, hitting the back of my head.[8]

Accidents and Abuses

Considering Roh's experiences in New York sweatshops, it is not surprising that there have been recent industrial accidents related to the sweatshop system. In Hamlet, North Carolina, in 1991, twenty-five workers died in a chicken processing plant that had been locked and had no fire alarms or sprinklers.

In August 1995, local and federal law enforcement agents conducted a raid on a sweatshop in El Monte, California, just east of Los Angeles. They learned that seventy-two immigrants from Thailand had been forced to work under slave-like conditions there. The workers were locked in an apartment complex surrounded by barbed wire. They made sixty-nine cents an hour and were threatened with rape and murder if they stopped working.

In July 1996, police raided four garment sweatshops in New York City. At one of these, workers were assembling a discount clothing line endorsed by model and actress Kathy Ireland. Ireland, in a statement released by K-Mart, her retailer, said that she would not tolerate the bad working conditions.[9]

The United States and many other countries continue to try to wipe out these illegal workplaces any way they can. In 1996, due to the increased interest in foreign and domestic sweatshops, President Clinton formed the White House Apparel Industry Partnership. Representatives from industry, labor, government, and public-interest groups got together to discuss how unions, government, and the apparel industry could work together to improve working conditions.

Estimates of the number of garment sweatshops in the United States today vary greatly. In 1996, the Department of Labor estimated that out of twenty-two thousand garment shops in the United States, at least half were in serious violation of wage and safety laws.

In April 1997, the National Labor Committee (NLC), a human rights organization based in New York, revealed that a line of clothing endorsed by then talk show host Kathie Lee Gifford was made in Honduran sweatshops. Much of the clothing was made by children who earned as little as thirty cents an hour. Gifford publicly expressed her outrage at this and she has since become an advocate of better working conditions for garment workers.

As recently as January 31, 2001, in a fire in a Manhattan building housing eight sweatshops, a garment worker was killed and several of his coworkers were injured.

An article in the August 8, 2001, issue of *Time* magazine announced the formation of a new coalition to fight sweatshops. "Despite years of public pressure against sweatshops, today's global retailers are greedier than ever, and more workers around the world are toiling in sweatshops to make their goods," said Bruce Raynor, president of UNITE, North America's largest apparel union. That union organized a coalition that includes the AFL-CIO (a federation of labor unions), religious groups, and United Students Against Sweatshops, which has chapters on more than two hundred college campuses.[10]

Winning the War Once Again

In the early and mid-1900s, workers built stronger unions and gained community support for their demands for safe and decent working conditions. The government adopted new labor laws that set minimum working standards. Today, organizing new workers into the union is a top priority for UNITE. This organization does not fight for workers only in the United States. It also fights for good jobs everywhere. UNITE supports workers in other countries who are fighting to organize their own unions to improve wages and working conditions because helping workers in other countries helps workers everywhere.

UNITE works with community leaders, religious leaders, students, and anyone who cares about ending sweatshops. UNITE also tries to hold politicians accountable for their actions on sweatshop issues.

An article at the UNITE Web site gives examples of some of the ways Americans (often unknowingly) support sweatshops in other countries. For example,

> Outside of Santo Domingo in the Dominican town of Villa Altagracia, 2,050 workers, mostly teenage girls or young women, make baseball caps bearing the names of America's great universities. Workers say that in a typical week at BJ&B they earn approximately $40 after 56 hours of work.
>
> Thousands of miles away, students, families, alumni and sports fans buy these caps at campus stores at Harvard, Rutgers, Georgetown, Cornell, Duke and other universities. They pay about $20 for a cap. The University makes about $1.50 from each cap from a licensing fee. Of that price, only 8¢ goes to the workers who made the cap. [11]

UNITE is working with college students across the United States in an effort to make sure colleges do not buy sweatshirts, caps, and other clothing with the school name from companies who use sweatshops to make these products. This is only one of the ways that the union is trying to eliminate sweatshops worldwide.

The Triangle Shirtwaist Company sign could be seen on the Asch building before the fire. Despite the inside of three floors being gutted by the fire, the building still stands today.

The Triangle Site Today

On March 25, 1993, eighty-two years to the day when the tragic Triangle fire took place, representatives of the National Park Service, the International Ladies' Garment Workers' Union, and the New York City Fire Department dedicated the site of the fire as a National Historic Landmark. The Triangle Shirtwaist Factory Building is located at 23-29 Washington Place in New York City. The property is now used as classrooms and offices by New York University. It is not open to the public.

The Last Survivors

Many of the survivors of the Triangle fire went on to live long, productive lives. Two of those who did were Bessie Cohen (then Bessie Gabrilowich) and Rose Freedman (then Rose Rosenfeld). Cohen was a nineteen-year-old seamstress at the Triangle factory when the fire occurred, while Freedman was just two days shy of her eighteenth birthday. Both women lived to be 107 years old. Cohen died in February 1999. Freedman died in February 2001.

Commemoration

March 25, 2002, marked the ninety-first anniversary of the tragic Triangle fire. Each year, UNITE joins with the New York City Fire Department to commemorate those who died and to rededicate themselves to the legacy of worker safety. On September 11, 2001, New York City was struck by a

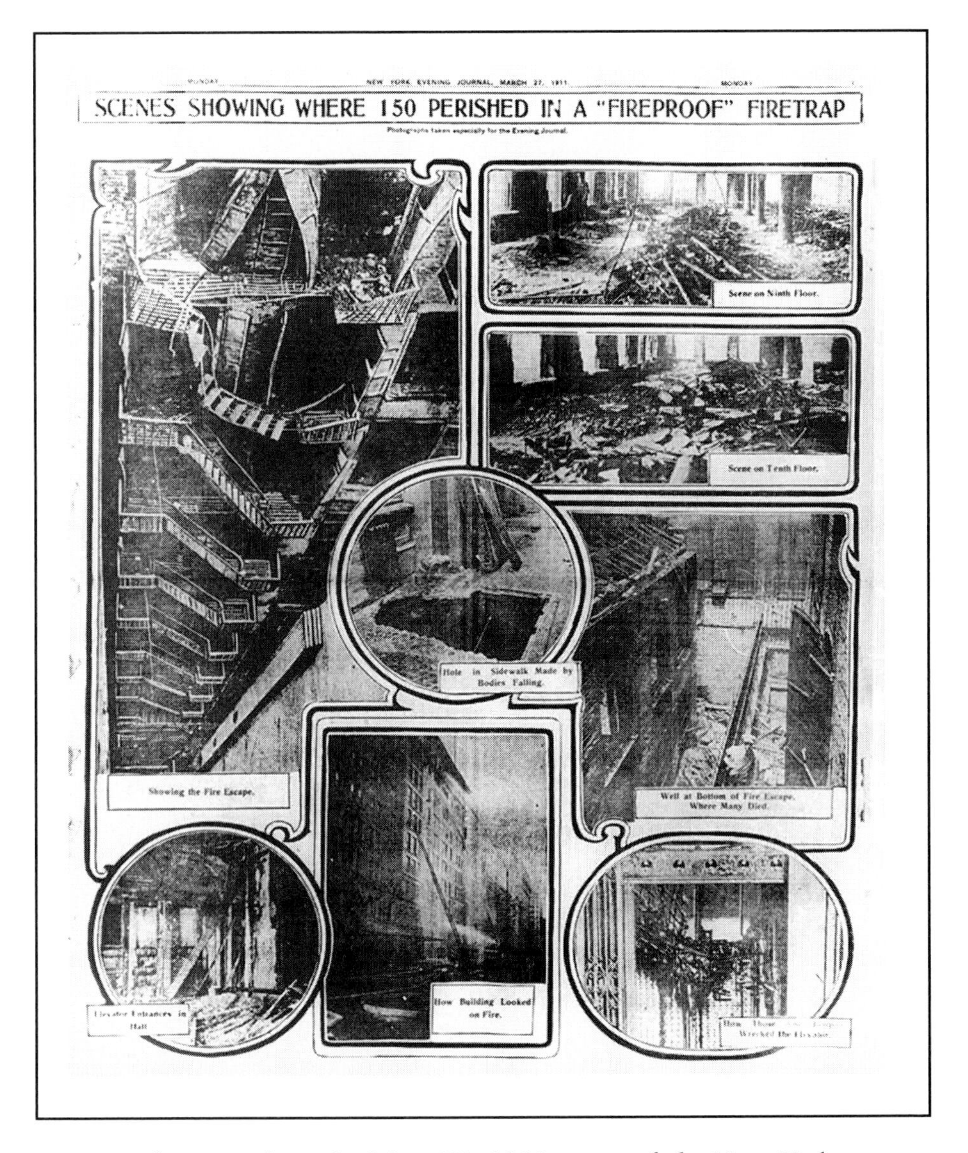

MONDAY NEW YORK EVENING JOURNAL, MARCH 27, 1911 MONDAY

SCENES SHOWING WHERE 150 PERISHED IN A "FIREPROOF" FIRETRAP

Photographs taken especially for the Evening Journal.

Scene on Ninth Floor.

Scene on Tenth Floor.

Showing the Fire Escape.

Hole in Sidewalk Made by Bodies Falling.

Well at Bottom of Fire Escape, Where Many Died.

Elevator Entrances in Hall

How Building Looked on Fire.

How Those [illegible] Wrecked the Elevator.

This page from the May 27, 1911, issue of the New York Evening Journal *shows the various places where bodies of victims were found.*

different kind of tragedy. Over three thousand people died in a terrorist attack on the twin towers of the World Trade Center. Most of those who died were working people. Hundreds were rescue workers—firefighters, police officers, and EMS workers. On Friday, March 22, 2002, Americans paid tribute to those who gave their lives in the heroic rescue effort of September 11, while they remembered those who died in 1911.

Legacy of the Fire

There will always be people who break the law and disregard rules and regulations regarding worker safety. However, the Triangle Factory fire made the city and state of New York, and later the whole country, realize that more detailed regulations and stricter enforcement of laws was necessary to be sure the most workers could be safe. The fire also made it very clear to more workers that they needed to join a strong union. As more and more members joined trade unions, these unions were able to become strong bargaining forces. Today, detailed regulations, strict enforcement of the laws, and strong unions seem just as important as ever.

Timeline

1840—Great famine in Europe causes the start of a wave of immigrants to the United States.

1890—Second wave of European immigrants begins.

1903—Women's Trade Union League (WTUL) is formed.

1905—International Ladies' Garment Workers' Union (ILGWU) is formed to protest low pay, fifteen-hour workdays, no benefits, and unsafe working conditions.

1909—Shirtwaist Makers go out on strike which is organized by the ILGWU and called "the Uprising of the 20,000."

1910—The cloakmakers' strike organized by the ILGWU.

1911—Fire at the Triangle Shirtwaist Factory kills 146 workers; Factory Investigating Commission is formed.

1912—WTUL is active in the investigation of the Triangle Shirtwaist Factory fire; Max Blanck and Isaac Harris, the Triangle Factory owners, are acquitted of any wrongdoing.

1913—The fifty-four-hour workweek becomes law in New York State; On March 11, Harris and Blanck settle with the twenty-three families who sued them. The two men pay damages of seventy-five dollars to each family.

1914—Thirty-six new laws are passed by this time, including a women's minimum wage.

1935—Congress passes the Wagner Act of 1935, which creates the National Labor Relations Board.

1938—*Life* magazine declares the war against sweatshops has been won.

1950—WTUL dissolves.

1960s—Sweatshops begin to reappear.

1970—Occupational Safety and Health Administration (OSHA) is created.

1991—Twenty-five workers die in a North Carolina chicken processing plant that is locked and has no fire alarms or sprinklers; In Thailand, 188 workers die in a fire, trapped behind locked doors.

1993—Asch Building becomes a National Historic Landmark.

1995—ILGWU merges with the Amalgamated Clothing and Textile Workers' Union to become the Union of Needletrades, Industrial and Textile Employees (UNITE).

1996—President Clinton forms the White House Apparel Industry Partnership.

1997—National Labor Committee reveals that a line of clothing endorsed by Kathie Lee Gifford was made in sweatshops in Honduras.

2001—In January, a fire in a Manhattan building housing eight sweatshops kills a garment worker and injuries several others; Rose Freedman, the last survivor of the Triangle fire dies in February; A new coalition to fight sweatshops is formed in August.

2002—Commemoration to mark the ninety-first anniversary of the Triangle fire and to also pay tribute to those who died in the September 11, 2001, terrorist attack.

Chapter Notes

Chapter 1. Fire in the Factory

1. Arthur E. McFarlane, "Fire and the Skyscraper: The Problem of Protecting the Workers in New York's Tower Factories," *McClure's*, Vol. XXXVII, September 1911, p. 474.

2. Ibid.

3. Leon Stein, *The Triangle Fire* (Ithaca, N.Y., and London: ILR Press, 2001), p. 37.

4. "Stories of Survivors: And Witnesses and Rescuers Tell What They Saw," *The New York Times*, March 26, 1911, <http://www.irl.cornell.edu/trianglefire/texts/newspaper/nyt_032611_1.html> (March 10, 2002).

5. "New York Fire Kills 148: Girl Victims Leap to Death from Factory," *Chicago Sunday Tribune*, March 26, 1911, p. 1.

6. "The Women of the Triangle Fire," *Revolutionary Worker #1046*, March 12, 2000, <http://www.rwor.org/a/v21/1040-049/1046/triangle.htm> (March 22, 2002).

7. Leon Stein, ed., *Out of the Sweatshop: The Struggle for Industrial Democracy* (New York: Quadrangle/New York Times Book Company, 1977), p. 188.

8. Stein, *The Triangle Fire*, p. 17.

9. "Stories of Survivors: And Witnesses and Rescuers Tell What They Saw."

10. Ibid.

Chapter 2. America's Factories in the Early 1900s

1. "Ellis Island History," *Ellis Island Immigration Museum*, 1998, <http://www.ellisisland.com/indexHistory.html> (February 3, 2001).

2. Edward T. Devine, *The Principles of Relief* (New York, Macmillan, 1904), pp. 35–36.

3. Annie S. Daniel, "The Wreck of the Home: How Wearing Apparel is Fashioned in the Tenements," *Charities 14, No. 1*, n.d., <http://www.tenant.net/Community/LES/wreck7.html> (March 23, 2001).

4. Leon Stein, ed., *Out of the Sweatshop: The Struggle for Industrial Democracy* (New York: Quadrangle/New York Times Book Company, 1977), p. 45.

5. Arthur E. McFarlane, "Fire and the Skyscraper: The Problem of Protecting the Workers in New York's Tower Factories," *McClure's*, Vol. XXXVII, September 1911, p. 468.

6. Stein, p. 66.

7. Pauline Newman, Joan Morrison, "Working for the Triangle Shirtwaist Factory," *History Matters*, n.d., <http://historymatters.gmu.edu/d/178/> (March 19, 2002).

8. Rose Cohen, *Out of the Shadow* (Ithaca, N.Y.: Cornell University Press, 1995), p. 113.

9. Miriam Finn Scott, "The Factory Girl's Danger," *The Outlook*, April 15, 1911, <http://www.ilr.cornell.edu/trianglefire/texts/newspaper/outlook_041511.html> (March 22, 2002).

10. Mary Van Kleeck, "Working Hours of Women in Factories," *Charities and Commons,* n.d., <http://www.tenant.net/Community/LES/hours10.html> (February 9, 2002).

11. Mary Van Kleeck, "The Shirtwaist Strike and Its Significance," *Mary Van Kleeck Papers, Box 29, Sophia Smith Collection, Smith College, Northampton, Mass., Women and Social Movements in the United States, 1775–1940*, n.d., <http://womhist.binghamton.edu/shirt/doc23.htm> (January 14, 2002).

12. Daniel, pp. 624–629.

13. Ibid.

14. Stein, p. 61.

15. "An Examination of the Role of Poor Women in the Sewing Machine Industry," *The Labor Frontier*, April 6, 1999, <http://home.wlu.edu/~metzgerc/technol/shirtwaist.html> (April 3, 2001).

Chapter 3. A General Strike Is Declared

1. "Labor Union," *Encarta Online*, n.d., <http://encarta.msn.com/find/Concise.asp?z=1&pg=2&ti=761553112> (March 9, 2002).

2. Constance D. Leupp, "The Shirtwaist Makers' Strike," *Women and Social Movements in the United States, 1775–1940*, n.d., <http://womhist.binghamton.edu/shirt/doc21.htm> (March 14, 2001).

3. Mary Van Kleeck, "The Shirtwaist Strike and Its Significance," *Mary Van Kleeck Papers, Box 29, Sophia Smith Collection, Smith College, Northampton, Mass., Women and Social Movements in the United States, 1775–1940*, n.d., <http://womhist.binghamton.edu/shirt/doc23.htm> (April 25, 2001).

4. Ibid.

5. Ibid.

6. Ibid.

7. "General Strike of the Ladies Waist Makers," *Women and Social Movements in the United States, 1775–1940*, n.d., <http://womhist.binghamton.edu/shirt/doc16.htm> (November 12, 2001).

8. "The Cooper Union Meeting," *The Call*, November 23, 1909, <http://www.ilr.cornell.edu/trianglefire/texts/stein_ootss/ootss_sg.html> (January 15, 2001).

9. Ibid.

10. Annelise Orleck, *Common Sense and a Little Fire* (Chapel Hill and London: The University of North Carolina Press, 1995), p. 60.

11. "Girl Strikers Tell the Rich Their Woes," *Women and Social Movements in the United States, 1775–1940*, n.d., <http://womhist.binghamton.edu/shirt/doc5.htm> (March 12, 2002).

12. Grace Potter, "Women Shirt-Waist Strikers Command Sympathy of Public," *Women and Social Movements in the United States, 1775–1940*, n.d., <http://womhist.binghamton.edu/shirt/doc3.htm> (November 2, 2001).

13. Ibid.

14. "Miss Morgan Aids Girl Waiststrikers," *Women and Social Movements in the United States, 1775–1940*, n.d., <http://womhist.binghamton.edu/shirt/doc4.htm> (November 2, 2001).

15. Mary Clark Barnes, "The Strike of the Shirtwaist Makers," *World Today*, March 1910, <http://web.gc.cuny.edu/ashp/heaven/text9.html> (March 12, 2001).

16. International Ladies' Garment Workers' Union: Let's Sing! Educational Department, "The Uprising of the Twenty Thousands," *The Triangle Factory Fire*, n.d., <http://www.ilr.cornell.edu/trianglefire/texts/songs/uprising.html> (July 24, 2002).

17. Constance D. Leupp, "The Shirtwaist Makers' Strike," *Women and Social Movements in the United States, 1775–1940*, n.d., <http://womhist.binghamton.edu/shirt/doc21.htm> (March 12, 2001).

Chapter 4. A City Mourns

1. "141 Men and Girls Die in Waist Factory Fire; Trapped High Up in Washington Place Building; Street Strewn with Bodies; Piles of Dead Inside," *The New York Times*, March 26, 1911, p. 1.

2. "New York, Episode Four: 1898-1918," *PBS Home Video*, 1999.

3. Leon Stein, ed., *Out of the Sweatshop: The Struggle for Industrial Democracy* (New York: Quadrangle/New York Times Book Company, 1977), p. 193.

4. "Rose Freedman," *Injured Workers' Alliance*, 1998, <http://www.injuredworker.org/Letters/Rose_Freedman.htm> (March 20, 2002).

5. Michael T. Kaufman, "Bessie Cohen, Survivor of 1911 Shirtwaist Fire, Dies," February 24, 1999, <http://www.ishipress.com/shirtwai.htm> (March 3, 2002).

6. "Echoes from the Triangle Fire," *The Ladies' Garment Worker*, September 1911, <http://www.ilr.cornell.edu/trianglefire/texts/newspaper/lgw0911.html> (March 12, 2001).

7. "Report of the Joint Relief Committee, Ladies' Waist & Dressmakers' Union," *No. 25 On the Triangle Fire Disaster*, January 15, 1913.

8. Martha Bensley Bruere, "What is to be Done?" *Life and Labor*, May 1911, <http://www.ilr.cornell.edu/trianglefire/texts/stein_ootss/ootss_mbb.html> (February 2, 2001).

9. Ibid.

10. Leon Stein, *The Triangle Fire* (Ithaca, N.Y., and London: ILR Press, 2001), p. 152.

11. Ibid., p. 155.

12. Ibid., p. 88.

Chapter 5. Investigations and a Trial

1. Leon Stein, *The Triangle Fire* (Ithaca, N.Y., and London: ILR Press, 2001), p. 135.

2. Ibid., p. 139.

3. "Blame Shifted on All Sides for Fire Horror," *The New York Times*, March 28, 1911, <http://www.ilr.cornell.edu/trianglefire/texts/newspaper/nyt_032811.html> (March 15, 2001).

4. Rose Schneiderman, "We Have Found You Wanting," *The Survey*, April 8, 1911, <http://www.ilr.cornell.edu/trianglefire/texts/stein_ootss/ootss_rs.html> (February 10, 2001).

5. "Echoes from the Triangle Fire," *The Ladies' Garment Worker*, September 1911, <http://www.ilr.cornell.edu/trianglefire/texts/newspaper/lgw0911.html> (March 12, 2001).

6. Factory Investigating Commission, *Preliminary Report* (Albany, New York: The Argus Company Printers, 1912), Vol. 1, p. 15.

7. Ibid., p. 20.

8. Testimony of Kate Alterman, "*People of the State of New York* vs. *Isaac Harris and Max Blanck*."

9. Ibid.

10. "147 Dead, Nobody Guilty," *Literary Digest*, January 6, 1912, <http://www.ilr.cornell.edu/trianglefire/texts/newspaper/ld_010612.html> (March 12, 2001).

11. "The Story of the Triangle Fire," March 2, 2002, <http://www.ilr.cornell.edu/trianglefire/narrative6.html> (March 10, 2001).

12. "147 Dead, Nobody Guilty."

Chapter 6. Steps to Improve Working Conditions

1. "Waist Factory Fire," *The New York Times*, March 26, 1911, p. 5.

2. Philip S. Foner, *Women and the American Labor Movement* (New York: The Free Press, 1979), p. 359.

3. "New York, Episode Four: 1898-1918," PBS Home Video, 1999.

4. "Between a Rock and a Hard Place," *American History Sweatshop Exhibit*, n.d., <http://americanhistory.si.edu/sweatshopshistory/2t125.htm> (March 12, 2001).

5. Robert J. S. Ross, "Sweatshop Police," *Behind the Label*, n.d., <http://www.behindthelabel.org/> (March 22, 2002).

6. Ibid.

7. "Sweatshops: Harsh Conditions Create Public Support for Reform," *Hearts & Minds*, February 2, 2001, <http://heartsandminds.org/articles/sweat.htm>(March 22, 2002).

8. "Sweatshops Timeline," *Undercover Diary*, MSNBC, 1998, <http://www.msnbc.com/onair/nbc/dateline/shop.asp> (March 23, 2002).

9. Ibid.

10. Steven Greenhouse, "Labor Groups Join Coalition to Eliminate Sweatshops," *The New York Times*, August, 8, 2001, <http://www.globalexchange.org/economy/corporations/sweatshops/nytimes080801.html> (November 19, 2001).

11. "Was Your School's Cap Made In This Sweatshop?" *UNITE!*, 2000, <http://www.uniteunion.org/sweatshops/schoolcap/schoolcap.html> (March 20, 2002).

Further Reading

Dash, Joan. *We Shall Not Be Moved: The Women's Factory Strike of 1909.* New York: Scholastic, Inc., 1996.

De Angelis, Gina. *The Triangle Shirtwaist Company Fire of 1911.* Philadelphia: Chelsea House Publishers, 2001.

Sherrow, Victoria. *The Triangle Factory Fire.* Brookfield, Conn.: The Millbrook Press, 1995.

Internet Addresses

"How Was the Relationship Between Workers and Allies Shaped by the Perceived Threat of Socialism in the New York City Shirtwaist Strike, 1909–1910?: Document List." *Women and Social Movements in the United States, 1775–2000.* © 1997–2002. <http://womhist.binghamton.edu/shirt/doclist.htm>

"Remembering Rose Freedman, last survivor of the Triangle Factory Fire." *CBC* © 2001. <http://radio.cbc.ca/programs/thismorning/sites/people/triangle_010225.html>

"The Triangle Factory Fire." *Kheel Center at Cornell University.* © 1998–2002. <http://www.ilr.cornell.edu/trianglefire/>

Index

SAUGUS PUBLIC LIBRARY
3 1729 00094 714 0

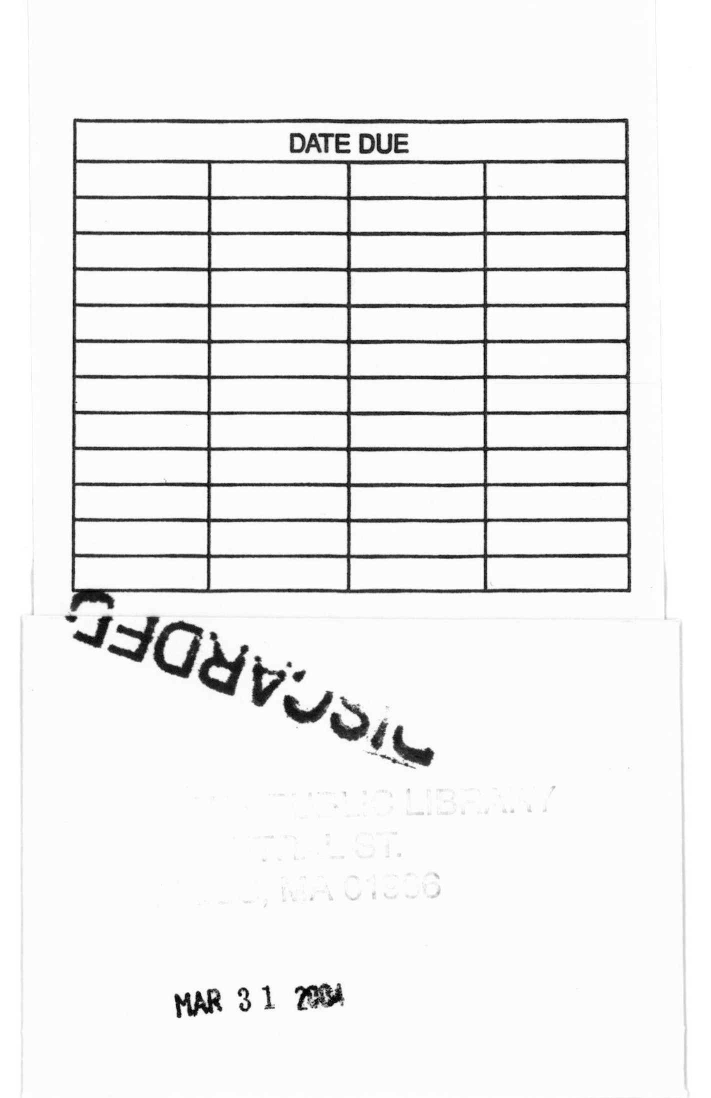
DATE DUE
DISCARDED
MAR 3 1 2004